CHALLENGER'S SHADOW

Did Government and Industry Management Kill Seven Astronauts?

By

John C. Macidull

And

Lester E. Blattner

Llumina Press

Second Printing

ISBN: 978-1-59526-730-6

Printed in the United States of America by Llumina Press

Library of Congress Control Number: 2007900059

DEDICATIONS

To my parents,
Dr. Joseph C. and Mrs. Mary Frances King Macidull
John C. Macidull

To those families who lost a son or daughter in the space program, and for those who value truth as the property of no individual, but the treasure of all.
Lester E. Blattner

PRECIS
Challenger's Shadow

By Macidull and Blattner

Of course, I said to Blattner, it is now known that the astronauts survived the January 28, 1986, explosion. The proof was kept from the Presidential Commission and the American public by NASA until after NASA had settled all claims (in the dark) with all of the astronauts' survivors. The details of those compensations are still classified. Then there is the question of the probable cause of the accident. The O-rings presented a convenient focus on inanimate objects that could not be brought into a court of law, rendering NASA unaccountable for its previously documented errant behavior.

Amid the conflicting motives of political, government, military, and commercial entities, top management at NASA and Morton Thiokol ignored established safety requirements. The Challenger was launched with known, critical rocket seal problems. It was also launched outside its environmental database envelope–a crucial consideration routinely given even to dogs, mice, and monkeys.

Since the Commission disbanded in 1986, having completed its *Report of the Presidential Commission on the Space Shuttle Challenger Accident*, much new information is seeing the light of day through this book. The Freedom of Information Act, the National Archives, and numerous interviews draw an inside picture of what actually happened before, during, and after the Challenger tragedy and subsequent investigation. Also, *Challenger's Shadow* includes new insights for both the aerospace professionals and lay people who have a large equity in the U.S. space program. Another purpose of this work is, we hope, to provide awareness for future engineers so this type of idiocy won't happen again–at least not in the near twenty-year future.

O-Rings

At this point it is important for me (Macidull) to describe what O-rings are and what they do, as they are instrumental to this history. They can range in size from the quarter-inch diameter rubber rings in your kitchen faucet and garden hose connections, up to hundreds of inches in diameter like those used in the Shuttle boosters. They are O-shaped rings of rubber-type material, and their job is to stop leaks. In the Shuttle boosters, they are like thick spaghetti that is placed in grooves. The best way to describe them is like large hula-hoops that are 12.16 feet (3.65 meters) in diameter. The Space Shuttle Challenger boosters had twenty-eight of them, each intended to keep hot gas and flame from leaking out when the booster rockets were firing. The Shuttle has two boosters, each consisting of six segments held together by six "field joints," plus another joint called a "nozzle joint" which swiveled at the bottom, for directional control. Each of these fourteen joints contained two of these rubber O-rings. Being of a rubber-like compound, they were susceptible to variations in temperature. The colder they became, the harder and less resilient they became, thereby inhibiting their ability to stop leaks. If these seals leak, flame could get to the fuel tanks, causing damage and a subsequent explosion. Even if there were no explosion, a leaking seal could deplete thrust from one of the boosters so they would be out of balance. If one of the boosters leaked enough on one side and there was not equal thrust from the other, then directional control would be lost, the Shuttle would spin, and there would be little hope of recovery. All people aboard and all equipment would then most likely be lost. This is why the subject of the Shuttle O-rings was recognized by nearly everyone with engineering training as being extremely important to consider before launch.

About the Authors

John C. (Jack) Macidull and Lester E. (Les) Blattner have fifty cumulative years of hands-on aircraft accident investigation and journalistic experience.

Jack Macidull graduated from the U.S. Naval Academy at Annapolis with majors in naval science and aeronautical engineering, and the University of Southern California with a Masters degree in public administration. After two tours piloting rescue helicopters in Vietnam, he first began his investigative career on military aircraft accidents while a jet aircraft test pilot for the U.S. Navy at the Naval Weapons Center, China Lake, California. He is also a retired Commander in the U.S. Naval Reserves, a fixed-wing aircraft carrier qualified pilot, and a recipient of six flight awards from four Secretaries of the Navy and two U.S. Presidents, among other awards. His relative thirty-year background and education includes employment as an aerospace engineer, a test pilot, a solid rocket engineer in research and development, and completing formal schools on the discipline of aircraft accident investigation. He investigated military and airliner accidents for the U.S. Navy, the National Transportation Safety Board and the FAA for over 20 years, during which time he was assigned to the staff of the Challenger Presidential Commission. He wrote the plan and some of the operational procedures for that investigation, as referenced in attachments 1 and 1-A of this work. He also organized and wrote much of that Presidential Commission's final report.

Lester Edward Blattner wrote books that helped define U.S. commercial aviation of the 1990s. These included McGraw Hill's *Aviation Intelligence Reports - U.S. Aviation and Europe*, 1992, *America's Airports* during the 1990s, and *U.S. Aviation and the Pacific Rim* in the 1990s. He served as Executive Editor for another McGraw Hill publication that explained the issues concerning privatization of America's airports. After serving in the U.S. Air Force, Les' career in aerospace began as a cub reporter for *Time Magazine* at Cape Canaveral a year before man went into space. He was present at the launching of America's original seven astronauts and wrote numerous articles and stories for *Time, Life* and *Fortune* magazines concerning the space program. It was while at Cape Canaveral that he interviewed such space luminaries as Dr. Werner von Braun and NASA's legendary space flight manager, Christopher Craft. Les left aerospace and went to work in Washington as a member of President Jimmy Carter's Administration, serving in numerous positions, including Director of Policy and Program Development. In the 1990s, Les wrote for the Federal Aviation Administration, Helicopters Association International, Airline Pilots Association, and the Flight Safety Foundation. He also appeared on national television as a communicator concerning air safety. His inclination toward aviation came about through his father, who was an original engineer at McDonnell Aircraft, where Les worked in the summer while attending the University of Missouri.

TABLE OF CONTENTS

PREFACE
By Macidull

Many books have been written on the Space Shuttle Challenger accident; however, none have been written about the actual accident investigation. This work attempts to explain the investigation, and in so doing lays bare much of what was heretofore unknown to the public concerning the accident and its causes.

On January 28, 1986, the Challenger spacecraft and seven crewmembers were lost. After only a few months, the media shifted its attention away from the tragedy; and many of the questions asked by the Presidential Commission were never answered. In my capacity as the only professional aircraft accident investigator on either the Commission or its staff, I lay blame on both NASA and Morton Thiokol management decision makers for setting the stage for this avoidable disaster.

Now, in 2002, sixteen years after the tragedy, information has become available to close this sad chapter in manned space flight. Blattner and I have obtained much of this historical information through the Freedom of Information Act (FOIA) requests and various interviews that we both conducted.

The American public paid for the space program but rarely had knowledge beforehand as to where or how their money was spent. The checks and balances inherent in our form of government were largely ignored in this tragedy and during the investigation. This "ignorance" contributed to the deaths of seven Americans and made us vulnerable regarding our national defense. This increased vulnerability lasted for nearly three years in a world of uncompromising beliefs by people who have little regard for life but great regard for their own beliefs.

This investigation was incomplete relative to the causes of the accident, largely due to lack of timely information. *Challenger's Shadow* attempts to present information in a way that may be useful in preventing similar occurrences. No other documentation, for example, has overlaid the crew dialogue during the launch in relation to what was happening outside the cockpit prior to the explosion. No document has addressed the inner workings of the Presidential Commission. No document has addressed the management decision process that precluded those experienced crewmembers from having a part in deciding their own fate. Above all, no document has addressed the human failures that caused this tragedy. O-rings, for example, were not the cause of the Challenger accident, but a result of the cause. Had NASA managers delayed launch until the ambient temperature reached the proven and recommended temperature (by Thiokol engineers to NASA of 53°F, or even the certified Thiokol qualification temperature of 40°F), this accident would most probably not have occurred. Instead, they decided to launch when the ambient temperature was 18° Fahrenheit. This was a temperature not only lower than anything previously experienced, documented or tested, but also a temperature that was below freezing, with a lot of obvious moisture all around. Did these decision-making managers actually not know that water begins expanding below 35°F, and that this is the reason that ice floats? Did they not know what problems this expansion could create with moisture extant, not to mention the other technical considerations relative to temperature and its untested effects on rubber rocket seals? Who were these people? Where on earth were they educated? And why on earth were they in positions of authority, making technical decisions concerning people's lives? *Challenger's Shadow* could be a case study for any business college that addresses management.

The O-ring temperature problem was well known long before the fateful Challenger launch. There had been nine O-ring failures, and problems during eight previous Shuttle launches, all at much higher and safer take-off temperatures (see Appendix IV) than on the day of this Challenger flight. Instead of fixing the problem, managers simply went against their own required safety regulations and formally tried to change those regulations so it would appear they were in compliance, in case anything later untoward happened. When two O-rings failed during the twenty-fifth Shuttle flight, which took place outside the known, tested, and previously certified environmental requirements of both NASA and Morton Thiokol, the "accident" was not a surprise to many either inside or outside NASA or Thiokol. Bob Ebeling, a Morton Thiokol engineering manager, made several relevant statements after the accident. Among those recorded: "There was another program manager who came by my office and said, *Bob, are you concerned about an 18 degree (F) launch? I said, what business has anyone got even to think about that. This is no-man's land. This is a big gray area."* During another meeting he said, *We discussed what might happen below our 40-degree qualification temperature. And, practically to a man, we all decided it was catastrophic* During a final meeting the day before the launch he said, *We recommended no-launch. The entire Thiokol group (of engineers) recommended no-launch.* But in this meeting he said, *Thiokol managers overruled their engineers and recommended to launch. NASA accepted Thiokol's recommendation to launch.* And later, also on MSNBC, Ebeling said the following: *I made a prayer. That prayer was a negative one ... Lord, make me and all these other engineers wrong. Let it go. And I thought (after liftoff) we were home free.*

The scapegoats for the accident were the O-rings. But this revelation hid the reason for the accident and focused public attention on hardware rather than human failings. Those responsible for the accident were never held responsible before the Presidential Commission or in civil court. The O-rings were perfect scapegoats, inanimate objects that conveniently could not testify. There were established safety procedures for O-ring problems developed over the years, the result of many decades of research by thousands of Americans into space flight. For managers to simply ignore these efforts and requirements constituted an insult to all of us.

The Challenger accident investigation was, and remains today in terms of people and money, the largest and most expensive accident investigation in history, involving over 3,000 people and costing more than $100 million. This book sees the tragedy as an example of human error and faulty decision making by government and industry managers, who then tried to cover up their lack of good or reasonable or humanly acceptable judgment by placing blame on hardware. It is a history about the subversion of due process by government managers with an agenda other than the one for which they were paid—to keep the public trust and perform as stated in their job descriptions. Also, *Challenger's Shadow* is about cover-ups and obfuscations in front of a presidential commission and about the lack of accountability of those in positions of authority.

After more than a decade of research, *Challenger's Shadow* provides facts, dialogues, views, and conclusions not previously available to the public. It covers the explosion, its aftermath, and the relative political, financial, legal, and management factors that combined to not only cause the accident but also to inhibit the government investigative process.

SHUTTLE COLUMBIA

By Macidull

The management problems with the Challenger 'accident' in 1986 paled to insignificance compared to the management problems seventeen years later with the Columbia 'accident'.

With Challenger, the crew had no time or choice to do anything after liftoff. With Columbia they had three options and over sixteen days to do something after liftoff. But they were given no information to make any decision concerning their lives, their crew's lives, or the spacecraft. There were no rear-view mirrors on the shuttle. All external visual information had to come from the ground. The first option was to abort after a large piece of 'foam' came off the external tank during liftoff and impacted the underside of the left wing and its protective heat tiles. The tiles were not needed to get into space, but were critical for returning into the earth atmosphere at high speed and hot air friction. And 'foam', that NASA liked to call it, was not 'foam' at that temperature and humidity. It was heavy.

Their second option was to inspect any damage that might have been done to the tiles under the left wing. But they, the crew, were not informed of any possible damage that could result in the loss of their lives. An inspection would have been fairly simple. Aside from high resolution ground cameras around the world that had been used on the shuttle for thirty years, the astronauts could have gone out on a tether and take some photos for ground analysis. But they were not informed of any pending problem.

If an inspection revealed the possibility of catastrophe, there was a third option left. The Shuttle Atlantis was set to go in five weeks. Without the mission load and computer diagnostic requirements, fueling and launch could have taken place in less than three days. On board the Atlantis would have been a repair crew. The Columbia crew would be simply transferred to the Atlantis and returned home. Hardly a big deal. The repair crew would fix the tiles. And if deemed impossible, the Atlantis would come back and pick them up in a few days. None of these above options were considered by 'management'.

And 'management' in NASA (except for John Young), had little piloting experience, and no education or background to be in charge of people's lives in such a highly technical and dangerous environment. They were basically rookies in charge because of political manifestations at the time.

###

CHAPTER I
Background -The Improbable Accident
By Macidull and Blattner

NASA efforts during the 1960s and 1970s resulted in a nearly seamless transition from manned research and development in space, to dependable, cost-effective space transportation. On April 12, 1981, the orbiter Columbia (STS1 Space Transportation System, flight number 1) blasted off from Cape Canaveral on a demonstration mission that ended successfully on April 14. NASA was back in space with a sophisticated manned spacecraft, ushering in a new era that the agency hoped would bring dependable, economic space travel.

Still, in the decade of the 1980s, it was not the Russians but rather the American public and ultimately Congress that put pressure on NASA. The agency was then begging for budget dollars. In fact, NASA's budget was directly related to the success of the new space Shuttle program. It was up to the agency to demonstrate to Congress that the nation was well served and that the Shuttle program budget was justified.

Yet, by 1986 something had changed. At first that *something* wasn't definable. Perhaps there had been a disconnect between technology and management, or a disconnect between science and human survival values. Astronauts, for example, used to have a say in their own business and destinies. The first seven, for example, insisted on having a window in their space capsule. Apollo astronauts, before going to the moon, would get together on the beach at Cape Canaveral and draw the orbital mechanics and equations of the paths of earth, moon, their lunar module, and mother ship. They made these drawings with sticks in the sand, not only to get it right in their heads but also to keep any written materials from getting into the hands of the Soviets. The waves, between known overhead Soviet spy satellite over-flights, washed these diagrams away—but not before they were imprinted and agreed to in the minds of Armstrong, Aldrin, and Collins.

As the details of this book will illustrate, the Challenger crew was kept out of the loop on a launch decision that would ultimately kill them. Because of automation and computers, they didn't have much to do during the launch except monitor instruments. Nearly everything was required to be computer operated. With thousands of mechanical operations each second, all requiring monitoring, computer-programmed automation was and still remains necessary during the flights of these later machines.

But somewhere along the line of required automation, the crew was relegated to the role of automatons. Regardless of their years of engineering, flight experience, and training, they were kept in the dark and not allowed to participate in those life-and-death decisions that were being made by managers who had no comparable backgrounds. And these Challenger pilots undoubtedly knew what was what. Pilot Mike Smith's close friend and college classmate, Rick Purnell, was watching the Super Bowl with Smith's brothers the night before the first scheduled launch (January 27, 1986). He called Mike to wish him luck. During the conversation, Mike said, according to Rick, *Get my brothers to go out and buy up all the anti-freeze they can get, because this mother is frozen up*! I asked if I could use this quote in this book. The answer was something like, *Absolutely, and it is about time that some of these questions were answered.*

There is no record, written, recorded, or even hearsay known to these authors that shows that any of the Challenger astronauts were allowed to participate in the decision to launch them that day. To help understand what happened, we have to go back in time and attempt to address the physical and personal circumstances of record around the Challenger launch.

On the morning of January 28, 1986, twenty minutes before the Challenger launch from Pad 39-B at Cape Canaveral, the outside air temperature had crept up from freezing to around 35°F. (This temperature was not the same as the local temperatures next to the liquid hydrogen tank and the O-rings, which were much lower. See Appendix IV.) And the Launch Control Center reported to the

Mission Management Team that ice remained on the left solid rocket motor (SRM) casing. Nevertheless, there were no delays in the launch, no further holds, and no further formal discussions about the cold weather and how it might affect the launch. The countdown continued.

At T plus 110.250 to 110.252 seconds, the Cape Canaveral Range Safety Officer activated the radio signals that destroyed the Solid Rocket Boosters (SRBs) in flight. According to high-resolution film from NASA camera #E202 during the active investigation, the crew capsule came out of the explosion intact and not rotating to any degree that would incapacitate the crew. *Post-commission*, analysis showed that it was slowly rotating at about two RPMs when coming out of the fireball. Also, it was determined that the umbilical wiring that used to connect the capsule electrically to the cargo section of the orbiter was trailing behind it like the tail of a kite, causing the crew capsule to stop rotating and stabilize in a forward direction before hitting the water. Wreckage analysis showed that the capsule impacted the ocean in a nose-down, slightly left-wing-down attitude (if there had been wings).

As the dive toward earth began, the crew probably went though their training procedure as explained in Chapter VI. The pilots aboard Challenger, without flight control (wings or tail), probably just watched as the vast Atlantic Ocean rushed toward them. It was over in two and a half minutes after the explosion, and they died in less than a second after impact with the ocean.

NASA kept the existence of conclusive physical wreckage evidence from the Presidential Commission. This evidence later established that the crew was not incapacitated by explosive decompression after the explosion. Macidull made requests through the Presidential Commission for NASA to provide such personal effects that may have been recovered, such as toothpaste tubes, medicine bottles, deodorant containers, and cameras to help establish positive proof of explosive decompression. This request was denied.

Six days after the Challenger tragedy, the accident investigation began by a presidential executive order. The process of choosing the members of this Presidential Commission remains largely unknown. The members chosen, however, are addressed in Chapter II of this book.

The tragedy was a dark time for NASA, an agency that depended on the good will of the public for support in congressional budget battles. But NASA management, as discussed in later chapters of this book, was involved heavily in impeding the investigation. As the later facts and analyses indicate, this was not only an improbable accident, but it was surrounded by two almost unimaginably incompetent management teams.

CHAPTER II
Macidull Inside The Presidential Commission
By Macidull

On February 3, 1986, President Ronald Reagan signed the executive order establishing the Presidential Commission on the Space Shuttle Challenger Accident. During the following two weeks, Commission Chairman William P. Rogers recognized that the Commission needed technical help with the investigation. He contacted then-Secretary of Transportation Elizabeth Dole. She referred him to FAA Administrator Admiral Donald Engen, who called me into his office. At that time I had worked for him for several years. He assigned me to help with the Commission on two conditions; one, *only* if I was interested; and two, the Commission was not to waste my time. If either became a problem, I had permission to leave. He said he couldn't afford to lose me on ongoing projects, mainly the FAA Safety Hotline and FAA System Safety Program, for make-work endeavors. I said I was pleased to help, not knowing what he meant by "make-work." So I reported to the Commission headquarters in Washington on Friday, February 21, 1986.

My recollection of impressions on that first day was one of a disorganized place. There were two secretaries, the assistant to the executive director of the Presidential Commission staff, a lawyer, an administration officer assigned by the Air Force, an astronaut—and no commissioners to be seen. That was it. Upon entering the offices I introduced myself to Major (USA) Tom Reinhardt, on loan from the Office of Management and Budget (OMB). He was on loan from the Army and had spent a decade as an aide to

Dr. Alton Keel, executive director of the Presidential Commission staff.

Reinhardt spent the next half-hour showing me newspaper clippings of Dr. Keel and his impressive record, talking about the Assistant Secretary of the Army and other titles, as well as his relationship with then-Secretary of State George Schultz. Finally, I asked to see a copy of the plan for the investigation. Reinhardt said there was no plan. I asked what the thirteen commissioners had accomplished after three weeks. Reinhardt replied that he didn't even know where most of the commissioners had been during the past week. I was then introduced to the rest of the Commission staff that was present—Astronaut John Fabian, William Dupree from the Defense Department's Inspector General's Office, Lieutenant Colonel Steven Hyle, an administrative officer from the U.S. Air Force, Pete Rustoven, a legal aide to Rogers, and Lorraine Walton, a NASA secretary.

Commission headquarters was not what one could call a well-oiled operation. Pete Rustoven was disgusted with it and said he was leaving the Commission the following week, which he did. Fabian told me what he knew, but it was light on substance; and he seemed to feel some of the same disgust as Rustoven.

It was a long day. That evening, I thought about the advice Admiral Engen had given me ... *"don't let them waste your time,"* and of Pete Rustoven's take on it. My experience at that time had included over 15 years of accident investigations. From what I had seen so far, the real question arose as to whether I should be involved with this Commission. I thought the effort was, so far, without professional technical guidance or direction. Most importantly, there was no plan—and hence, no delegation of resources, no accountability, no documentation order or control, and there was not a single person on either the Commission or its staff with any experience in aircraft accident investigation. I decided that I would not stay unless there was a workable plan for a reasonable investigation—so I spent the weekend writing one (Appendix I). Also, some of this motivation was loyalty to my U.S. Naval Academy classmate Mike Smith, who probably would have wanted someone from his school to look into his death.

At Commission headquarters on Monday, I introduced myself to Chairman Rogers' personal secretary, Marie Hunter, who then introduced me to several others. I felt awkward as I said something like, "Hi, I'm Jack Macidull, and here is a plan for this investigation."

Both Chairman Rogers and Vice Chairman Neil Armstrong seemed pleased, perhaps relieved, that a draft plan had been done. Later, Astronaut Sally Ride pulled me aside and thanked me. I cautioned her that it was a plan written by someone who knew little about NASA or space flight, but I did know something of technical aircraft investigations. She replied, "Well, at least it's a plan!"

During that week Rogers started implementing the plan. At a meeting of the commissioners, with the chairman presiding, Rogers said that assignments for the specific commissioners were in specific areas, or panels, of the investigation plan. Media relations support was to be provided by Mark Weinberg from the White House as well as by Herb Hetu, Rogers' private consultant.

Assignments were made to the panels. Several of the commissioners wanted to be assigned to two panels, which was no problem for Rogers. When Commission assignments were posted, this is how the Commission was structured:

Development and Production: Commissioners Joseph Sutter, Dr. Albert Wheelon, Dr. Arthur Walker, and Robert Rummel. Staff support was provided by William Dupree and John Hungerford.

Mission Planning and Operations: Commissioners Dr. and Astronaut Sally Ride, Robert Rummel, Robert Hotz, and David Acheson. Astronaut John Fabian provided staff support.

Pre-Launch Activities: Commissioners David Acheson, Dr. Sally Ride, Dr. Richard Feynman, Dr. Arthur Walker, and Dr. Albert Wheelon. Maj. John Chase, USMC, and Astronaut and Lt. Col. Brewster Shaw, USAF provided staff support.

Accident Analysis: Commissioners Neil Armstrong, Major General Donald Kutyna, USAF, Dr. Eugene Covert, and Dr. Richard Feynman. Staff support was provided by me, FAA investigator and Commander, USNR, and Ron Waite, consultant to Neil Armstrong, who had a long and valuable history with NASA as director of operations at the test facility of Edwards Air Force Base at Muroc Dry Lake Bed in California. According to Armstrong, Ron knew where the skeletons were buried, and he would "take no prisoners." He was right. Ron's knife-edged diplomacy in dealing with some people in high places saved the Commission and me much wasted effort.

And so did Randy Kehrli, a lawyer on loan from the Department of Justice. He sat next to me in the office and did a great deal to screen out people who wasted our time. As a national tragedy with worldwide attention, there were many who just wanted to walk in, absorb the atmosphere, and satisfy their curiosity and ego needs. For example, I wasted half an afternoon talking to a representative from Senator Pell's office who provided a constituent letter explaining that, during a séance, it was found that the explosion was caused by a device on the Shuttle set off by a radio signal from a hand-held communicator carried by someone in the civilian observation area during the launch. There was no proof or evidence of any kind. The only thing I could think of was a quote from Abraham Lincoln who, when asked what he thought about someone's speculation that had nothing to do with reality, responded, "Well, for those who like that sort of thing, I suppose that is just about the sort of thing they would like."

Kehrli also provided many of the applicable legal questions for the public hearings and participated in many of them. Unfortunately, he was never given credit for his efforts, experience, education or participation. This exemplified the small-mindedness of politics in Washington. The presidential report had to include Kehrli's questions, which often addressed the heart of the matter. The transcripts went to final print with everyone's name properly included. But when the report was printed, after revisions were no longer possible, Kehrli's name alone had been replaced. The report failed to identify him as the originator of every single one of his erudite and useful questions. Instead, his name was replaced by the word "Question," which would imply to the reader that the previous person named in the transcript had asked the question. This was an example of the puerile attitude present at the time, with some of those in positions of power who could not keep from attempting to either steal or eliminate due credit from others.

At about this time Chairman Rogers brought in four FBI agents. "Gum-shoes," he called them—people who performed over 200 interviews. Some of us wondered why they were talking to the NASA engineers in "street talk" while we were doing the same thing, only in engineering terms. But no harm was done. Perhaps the Chairman wanted to cover a base of which we were unaware.

Logistical arrangements, travel reimbursements, and typing were spelled out. NASA would provide secretarial support. Marie Hunter provided additional secretarial assistance for Chairman Rogers, and Marty Black from the White House did the same for Vice Chairman Neil Armstrong.

Again it became obvious that no one else in or around the Commission knew much about the discipline of investigating an accident involving human beings, with the possible exception of Astronaut Armstrong. But even his experience was limited to military and NASA in-house investigations. Then too, there was the time pressure. Most major aircraft investigations take more than a year and some take many years. As mentioned, in terms of money and resources, the Challenger investigation was larger than any accident investigation in history. Yet the Commission was given only four months to investigate and complete its report.

I requested that the Commission employ three more people on the staff with expertise in the areas of human factors, meteorology, and metallurgy. This turned out to be a hard-won fight before I finally got human factors expert Curt Graeber from NASA Ames and metallurgist Mike Marx from NTSB. I lost the fight to get Greg Salitolo, an expert in meteorology, Chief Meteorologist at NTSB, with whom I had worked for several years. However, he was gracious enough to provide information that eliminated wind shear as a contributing factor to the accident. There was strong jet stream activity that day, and other data showed that the rocket nozzles slewed nearly to their limits to compensate and maintain desired flight direction and flight path. But all wind encountered was within the design limits of the Shuttle system, and the system compensated for the wind in accordance with what the hardware was designed to do, without reaching its limits, without compromising safety, and without cause for concern.

During the next two weeks, after traveling to Cape Canaveral, my requests for substantial technical documentation from NASA engineers met with good success. They sent everything that I wanted, directly to the Commission. The flood of information from NASA engineers (they all, in my experience, only wanted to help in any way they could) became so voluminous that to just manage it became a challenge. But NASA management stepped in and impeded the flow of information. (Refer to Chapter VI for a more detailed discussion.)

The solution to document management was twofold. I requested ten large open bookcases that were placed in the bay area at headquarters. All the material was put in notebooks, labeled on the spine for instant identification, and placed in specific categories on the shelves by John Dunbar from CACI, a civilian company that specialized in large documentation organization. Initially CACI was contracted to have all documentation labeled and categorized in accordance with National Archive requirements for later storage. But for our immediate purposes the bookcases, notebooks, and categorizations, with the additional help of Steve Croll (also of CACI), greatly facilitated the investigation and enabled "cross-pollination" among the four panels, whose functions often lapped over into another panel's area.

About five weeks after the Commission was organized, it had five secretaries, a receptionist, two travel coordinators, two administrative officers, five documentation support specialists, two lawyers, four FBI agents, one professional aircraft accident investigator, and two media relations people. Nearly all of these people were needed and appreciated, and rarely did any have an idle moment. Offices were provided for commissioners, and desks with partitions were set up for staff in a large bay area.

NASA (with taxpayer money that could now not be spent on much else) also paid for the office rent, copy machines, secretaries, office supplies, commercial travel, contractors for documentation needs, and the final report.

Commissioners served without pay but received per-diem compensation for room and board while away from home. Most staffers' salaries were paid by their previous organizations' budgets. My Notice of Personnel Action of 02-23-86, in the block titled *Name and Location of Employing Office*, said, NATIONAL AERONAUTICS & SPACE ADMIN, WASHINGTON D.C. I still don't know where my salary came from during that period on the Commission staff, since my paychecks still originated from the FAA. Most staff personnel, 32 then, were made available by other government agencies and entities, which included the White

House, OMB, DOD, USA, USN, USMA, DOJ, USAF, USAFA, FAA, FBI, GPO, NASA, and NTSB.

Eight writing support people came in six weeks before the report was due. They were of the various disciplines of editor, writer, word processing supervisor, layout artist, printer, and designer. Most proved extremely valuable but could not write the report.

Also, about six weeks before the report was due, another critical staff group was formed that was called the Independent Test Observers. Six people from the USAF, TRW Aerospace Corporation, and NTSB were charged with witnessing and documenting SRM (solid rocket motor) propellant test firings. There are six propellant segments in each SRM. Each batch of mixed propellant is divided in half, one half for the left SRM on that launch, and one half for the corresponding right segment. Small propellant samples, about the size of a paint can, are taken from each of the 12 segments. Then, each corresponding pair of samples is burned to determine that they burn at the same rate. If not, then a mistake was evident. It is an almost foolproof check that this critical element in the launch process is safe. If the two solid rocket motors that help propel the Shuttle into space do not burn out almost simultaneously, catastrophe would result. This test observation group, in effect, eliminated this possibility as a cause of the accident.

The four Commission panels met with Rogers or Armstrong at least once a week to discuss their focus and travel plans. This way, Rogers knew what everyone was doing most of the time. When Rogers or Keel approved travel, the staff needed only to tell one of the secretaries when and where they were going. Usually the next day the requested airline tickets were on the desk. Much of the travel by the Accident Analysis panel was provided by the Air Force through General Kutyna. He scheduled trips out of Andrews AFB to the Cape, Huntsville (NASA Marshall), and Ogden, Utah (Morton Thiokol).

Then came the report to the President. I knew that it had to be done, and Armstrong told me that someone had to start writing the report. I agreed that it was a good idea and that is was about time that we got started on it. But he also related how Dr. Keel had brought in English professors from military academies to do just that, and he added that these teachers didn't know much about the accident, couldn't write the report, and that *I* should start writing it. Dr. Keel had mentioned several times during staff meetings that he had not made up his mind whether or not the staff would be included in the Presidential report. This is after the staff had done most of the investigation. I wrote up a format page with Dr. Keel's name at the top, secretaries next, and all us investigators below and took it to him, saying in my recollection, "Here is a format with names to include the staff in the report. If you don't like it, or have any suggestions, good luck on finding anyone to write the report." Here was a person telling me that those who did most of the work, who would then evaluate that work, and then write the work, were possibly not to receive any credit by even a mere mention of their names for the work they did. Sorry Charlie, was my attitude. There would be no report written by me unless all our names were in the report. Was this too much to ask? I didn't think so, then or now.

The English teacher assigned to me the next day was Captain William Bauman, USAF. I had assembled a pile of 26 notebooks containing all of the NASA test results concerning the accident, most requested by Armstrong and me. These tests took nearly 2,000 people three months to perform and document. I asked Captain Bauman if he was picked to write the report. His answer was "That's right." So I said, pointing to the stack of material, "Well, here's your research. Write the report."

The following day, Emily Trapnell informed me after her conversation with Dr. Keel that the names of the staff would be included in the final report. Later that day, Captain Bauman appeared at my desk to say he couldn't write the report because he was not technical and knew little about the investigation. I asked if he was willing to assist me. Bauman said he was ready to help in any way. So I asked him to tape together 20 feet of yellow legal paper and put the taped pages on the long conference table in the Commission conference room and to get a pair of scissors. In three days we had many words from NASA reports, Commission findings, and technical facts cut and pasted onto the yellow sheets. Then I started filling in the blanks from my notes and memory to help bind it together. This document was then rolled up like a scroll and sent to the secretaries in the writing room. They put it on computer disk and printed it out for Commission review. In the end, I thought that Captain Bauman proved very valuable and instrumental to the writing effort.

Also, the West Point writer, Colonel Anthony E. Hartle, assigned to a different panel, did a wonderful job. I watched him closely after his first draft became available. Knowing little of his subject at first, he spent, in my observation, about 80 percent of his time researching his subject and 20 percent writing. That explained, to me, the high quality of his work.

Shortly after the first draft was finished, Rogers called for a meeting. I sat on a couch in front of his desk. Several commissioners were standing behind him. Rogers told me that I had addressed components in the draft that had nothing to do with the accident. I responded, "Nine things, eight of which had nothing to do with the accident." Rogers asked me what these components were. He wrote them down as I mentioned them. Then he counted them off: "External Tank, Main Engines, Orbiter, Payload, Payload Interfaces, Satellite Inertial Upper Stage, Upper Stage Support Equipment, Solid Rocket Boosters, and Right Solid Rocket Motor [which contained the O-rings in question]." Rogers asked what else. I said that there should be nine items. Rogers counted down his list and asked me what my thinking was. I explained that the logic of it is to systematically address, evaluate and eliminate every major physical component of the Challenger "stack" as a technical cause or contributing factor to the accident in order to identify the true physical component of cause—and that this approach would result in a thoroughness that could be appreciated by many speculators, naysayers, and critics in the future. He agreed.

Part of the process of putting the report together was to review galley proofs, those that come prior to a final print. These proofs are in the exact layout and print the book will have, allowing the reader to see what the book will look like, page breaks and all, and make final corrections prior to final publication. They are expensive. Normally, the Government Printing Office (GPO) only provides one. In this case, because of the time constraints, Rogers insisted on every draft being of galley quality. He wanted to know what it would look like in final print with each draft. This turned out to be a good decision. After 21 galley proofs, the Commission and staff got a handle on how it would best fit together, which enhanced the effort relative to the time restriction.

A few days before each public hearing, Rogers met with various staff members and had them write down important questions. This method proved to be useful but somewhat irrelevant in conducting the 24 hearings between March and May 1986. This brings us to the two final points in this chapter: First, did the Commission establish the most probable cause of the accident? Second, what exactly was this Commission?

The report related the truth about the O-rings, but not about the actual cause of the accident in the opinion of this investigator.

The Commission: Who were these people? Granted, there is no process or procedure that restricts the President of the United States in his choice of personnel for his commissions. He can pick adolescent space buffs to accomplish a spacecraft investigation if he so desires.

None of the commissioners, or the Executive Director of the Presidential Commission Staff, had any professional experience as aircraft accident investigators. This is not to say that there were not other important roles to be filled. Early on, I started wondering about the commissioners, about why they were chosen, their qualifications, and later, what contributions they had made to the effort. Following are my short observations of six commissioners and the Executive Director of the staff, individually, by why I think they were chosen, their backgrounds, and what they contributed. Then I address the other seven commissioners.

Chairman William P. Rogers—Chosen because of his political and legal background as U.S. Secretary of State and U.S. Attorney General with excellent managerial skills in high-level organizations. *Contribution*: Made the whole thing work well. Given the time limitations, vast amounts of information, large egos to contend with, and friction within NASA, he proved vital to the effort, although I disagreed with some of the conclusions in the final report.

Vice Chairman Neil A. Armstrong—Chosen politically because of his unique reputation as the first person on the moon, earned respect, and knowledge of NASA. Also, he was the only commissioner with a firm, or in this case, *any*, technical grasp of the environment of space, space machinery, piloting space vehicles, atmospheric flight, aeronautical and aerospace engineering, and orbital mechanics. *Contribution*: Worked well with everyone. Kept technical meetings with NASA focused. Established early testing criteria, as well as tests for NASA—allowing it to accomplish detailed evaluations of parts such as O-rings. Helped materially with the editing and writing of the report.

Dr. Eugene E. Covert—He was chosen because he helped design the Shuttle main engines. He told me this after I asked him. I was suspicious about his knowledge of aeronautics, even though he was the head of the Aeronautics and Astronautics Department at MIT. Later I found that he did not have a degree in either aeronautics or astronautics, or if he did, it was not mentioned in his bio, which he wrote for the Commission report. Does a Doctorate of Science include aeronautics or astronautics? Curious. If not, then why was Mr. Covert, with no relevant degrees, in charge of education in the fields of aeronautics and astronautics at one of the country's most prestigious colleges of both aeronautics and astronautics? *Contribution*: He helped establish that the Shuttle main engines (MEs) operated perfectly, automatically shut down as designed during the explosion when overheating, and were not a cause or contributing factor to the accident. I inspected these engines closely and agreed with his evaluation. They had obviously suffered burn-through from the high oxygen input. This happened after the long lines from the upper oxygen tank fed the oxidant, during microseconds, into the engines after the liquid hydrogen tank came unzipped, preventing it from supplying enough fuel for burning within the temperature limits of the engine casings.

Robert B. Hotz—Chosen because he had been the editor-in-chief of *Aviation Week & Space Technology* magazine for 27 years, and probably knew nearly everyone under the sun in the field of space technology. *Contribution*: Helped on parts of the report, although he did little writing or editing himself. Brought in staff, one of whom (Lawrence J. Herb) proved extremely valuable in doing the photographic layout.

Major General Donald J. Kutyna, USAF—Chosen because he had been in charge of an unmanned Air Force missile accident investigation and because he did his graduate work at MIT where he knew Dr. Covert. *Contribution*: Provided Air Force transportation for commissioners and staff when required, on short notice. He proved extremely valuable with his hard work and attention to detail during the final writing and diagram organization in the report.

Dr. Sally K. Ride—Chosen because she was the first American woman in space and held a Doctorate in Physics from Stanford University. *Contribution*: Provided excellent and insightful preliminary questions and comments during Commission public hearings.

Dr. Alton G. Keel, Jr., Executive Director of the Presidential Commission Staff—Chosen because of his impressive background with various government organizations and his doctorates in engineering and physics. *Contribution*: Provided much organization and solved office procedural and equipment needs. Also approved the hiring on the staff of two people with critical investigation disciplines: human factors and metallurgy. Also, effectively coordinated work with the Government Printing Office during the process of report publication.

Other Commissioners

The other seven commissioners had truly unique and honorable backgrounds, but I could detect no contribution from them relative to the research, investigation, planning, or writing. Six had zero impact on the effort. One never showed up for any meetings after the first one; however, he sent four copies of his signature from four different pens for the report sign-off four months later. One was less than worthless, and we all would have been better off if he had not been around. He constantly tried to take credit for the Commission's work by talking to the press, against the Chairman's guidelines. At the time there were many occasions when I was asked what I thought of this particular person. My comment was usually the same—that the "strong forces" that hold protons together inside the nucleus of an atom pale to insignificance when compared to the force that attracts this person's mouth to the nearest microphone. When there wasn't a microphone around, he often screamed at secretaries for not immediately providing him with something. Most of us just thought he was sick or under some kind of medication, which he may have been. Strangely, he still has some reputation of "discovering" the O-ring problem. He did nothing of the kind, though he was a genius at self-promotion. If he had not shown up for anything, he would have then been only worthless instead of less-than, which would have been a boon at the time.

So we had 14 people in the Commission hierarchy—13 commissioners and an executive director of the small staff. Eight of these commissioners referenced their educations in their short biographies in the presidential report. Five had doctoral degrees in physics; three had degrees from MIT; two had taught or had been educated at Stanford. Aside from Armstrong, were there any degrees in aeronautics, aeronautical engineering, or aerospace? None. Any from schools like Embry Riddle, or similar schools like those in the armed services? None. Where there any from the NTSB Department of Aircraft Accident Investigation, or the FAA investigative counterpart, Office of Aviation Safety? None.

It seemed from the selection of commissioners that the high political ground was quickly attained by the type of club mentality that focused on similar educations and acquaintances, no matter whether their fields were much related to the effort. Virtually all aerospace or aeronautical degrees, for example, require comprehensive courses in physics first; but no physics degrees require similar courses in aerospace or aeronautical engineering. Five of these fourteen were physicists, one theoretical, and none operational or applied. Why, aside from adding some form of legitimacy to the Commission through famous name recognition for political purposes, were many of these people there, sitting at meetings and unable to contribute anything?

The Commission report came down to this: The O-ring was the weak link in the Challenger accident. But it did not emphasize that the Challenger should not have been allowed to fly after the first failure. According to initial requirements, the solid rocket motor O-ring design would have been discarded by NASA and fixed long before the final Challenger stack was even on the launch pad. And it should have been replaced with a newly designed gas-containment device that was tested and found to work in all temperatures (of record) at the launch site, as initially required through NASA's safety programs. The Commission report also said that NASA and Thiokol managers had blundered in their final decisions to launch Challenger. Yet there remained no accountability.

In reading the report today, it still begs the same question of who was responsible, and why no one was accountable. Seven people were dead, and our country had suffered other major losses in terms of training and experience to the nation, national security, and monetary investment. In the discipline of aircraft accident investigation, there is a category of probable cause labeled "Act of God." An example of this is when it is clear from recorded data that an aircraft was struck by lightning and crashed. During my years of experience, this probable cause had been used only once on a Boeing 747 over Spain. O-rings were not an "Act of God." The Commission did not say specifically who on NASA's management launch team, or management at Thiokol, could have contributed to sending Challenger on its way to that disaster. They identified several people during public hearings but did no follow-up. The Commission was provided questions by the staff for the public hearings; but when they (the Commission) received a bogus answer, the commissioners simply moved on to the next question. Why? Because none knew enough to ask follow-up questions, much less recognize when an answer was technically less than forthright.

Producing the Presidential Report was a difficult and sometimes a seemingly impossible job. However, this much is obvious: the report was a professionally prepared document with considerable research underpinning certain facts. Few people facing this responsibility could have produced a better document and service in only 120 days. The voluminous amount of material generated in only a few short months was indeed remarkable. But combing through the report pages years later, away from the glare of publicity and the pressure facing the Commission in its deliberations, I wonder why the commissioners not only failed to ask better questions but also failed to indict people for lying to the Commission. Was the lack of a full report to the American public the fault of an eclectic group of commissioners?

On the Commission were a former Secretary of State and Attorney General, Astronauts, a prominent lawyer, a Ph.D. professor in Science, a Nobel prize winner in theoretical physics, the former editor-in-chief of *Aviation Week and Space Technology*, an Air Force general, an aerospace executive, an astronomer/physics professor and associate dean at Stanford, a vice president of an aerospace firm, a skilled and highly rated pilot, and a doctor in engineering and physics. Collectively, this highly prestigious Commission could probably have run a country. Yet, my notes on the Commission from that time indicate that only six, or less than half the total, actually contributed to the investigation. Of those six, only two would ask tough questions, the Chairman and the first American woman in space.

Perhaps it is unfair to judge such highly skilled and dedicated professionals who, after all, were chosen and didn't volunteer (as far as we know) to represent the President of the United States, and thereby represent us. And the American people had a right to ask for the truth about what they saw on television on the morning of January 28, 1986. We had paid for that privilege, not only through taxes but also through electing a president to the White House and representatives to Congress. We did not want tunnel vision, political favor, or only technical answers that could not be deciphered in lay terms.

Of course, the burden of writing the Commission's final report did not fall on the commissioners. It fell on the staff. In such matters in Washington as elsewhere, the staff is usually the foot soldiers, the grunts that are asked to take that next hill. To me and to the others, the investigation was a war with a 120-day attrition rate. Though many good things resulted from this investigation, it still would have been, in my estimation, extremely valuable to have available to the staff at least one *New York Times, Washington Post,* or other investigative reporter who made his or her living by asking questions—someone skilled in seeing through cover-up, misdirection, and lies, such as some presented in Commission testimony and depositions.

A point that seldom appeared in print was the fact that the Commission was paid for by NASA from its FY86 budget. The irony is that the Commission was investigating NASA, who was not only paying for the Commission (with tax dollars) but was also providing its support staff. It took a lot of valuable investigative time just "keep the foxes out of the chicken coop."

Thiokol came out in the report better than it went into the Commission public hearings. NASA, as an institution, wasn't touched. NASA managers were slapped on the wrist, a small consequence, and quickly sought retirement. The ultimate paradox was that the Commission report was exemplary in effort and intent. However, it failed to deal with the human factors aspect of who was responsible for the launch decision.

The whole purpose of our legal system is to prevent future infractions of rules and laws in order to reduce the repetition of harmful actions. Who is to say in our highly technical world that there will not be another O-ring type of known danger? But to fix that O-ring, finally, after it had been known to fail, only *after* it failed catastrophically, solves little. Decision makers in the future, who are paid in the public trust but who are not held responsible for abiding by their public trust regulations—which are laws and required job descriptions—will not prevent future similar occurrences. The people involved in this launch decision were required to adhere to regulations that were developed over decades, by many thousands of Americans who learned how to successfully launch manned spacecraft. When these managers didn't adhere to those plainly detailed requirements and were then not held accountable, the underlying accident cause was neither addressed nor solved.

Yes, the O-ring part of the problem was physically corrected. And other concerns were dealt with beyond the recommendations of the Commission, like equipping the astronauts with parachutes. But there came no recommendations relative to management problems. NASA management today is very safety conscious, just like they were during the Gemini, Mercury, and Apollo programs. No one was lost on a flight to or in space during those decades. But the Shuttle program seemed to follow a 20–30 year cycle extant in many large, technical organizations. This cycle consists of new managers bubbling up in the organization through tenure or club-mentality who do not have the education and experience of their predecessors. When they get there, they bring a different mentality, one of little appreciation or hands-on knowledge and experience of engineering and thus of engineers. Their priorities become budget and profit and those things that immediately support budget and profit, such as scheduling and public relations.

This attitude is not life-threatening with organizations that make computer chips or coffee pots, where the marketplace and competition rule. The Shuttle program managers, in collaboration with the U.S. Air Force, designed the Shuttle to have no competitors. NASA obtained an "exclusive" on the market with Shuttle civilian research programs. In exchange, the Air Force got an exclusive on military satellites. Once this trap door was shut, the only source of funding, Congress, had no real options to consider except shutdown of the program and reinvestment in previously proven technology within all the services.

Today, NASA managers have put safety back on the top of the list of priorities, but what about ten to twenty years from now? Will budgeters and public relations people replace the safety-conscious NASA managers of today, after decades of safe space flight, again?

What If There Had Been No Commission?

What most likely would have happened had a presidential commission not investigated the accident? It is easy to assume that the O-ring problem would have been fixed and swept away from public memory. But there would have been no public hearings, or threats to subpoena information from NASA. Yet as it was, the short Commission existence made such instruments of the legal process nearly toothless. Further, the announcement by NASA that the crew died instantly in the explosion would not have changed. Most of all, today's Shuttle crews still might not be wearing parachutes during launch and recovery operations. As a result of the Commission hearings, top NASA managers quickly resigned and disappeared from public view. Had the investigation not taken place, these men would most likely have remained highly touted members of the elite NASA club (See Chapter III.) However, because of these considerations, and the requirement then for a presidential commission to accomplish any form of investigation, it was recognized that space flight is a form of transportation. Therefore, since the accident, and as a direct result of the Challenger investigation, the charter of the NTSB now includes space flight accidents.

After all these years, astronaut survival is a relative question. Of course we knew then that the Challenger astronauts did not survive the day. However, I knew that they *did* survive the explosion and recommended in a document recorded in the National Archives that in the future they be provided with parachutes. And we know that today's astronauts have parachutes and redundant systems engineered into their flight hardware. We also know that, once again, cost and schedule are secondary to safety. So in these ways it can be said that the lives of the Challenger crew were not lost in vain. Therefore, imperfect though it was, the Commission has had a beneficial impact on the U.S. space program.

The following chapter mentions a number of personnel from NASA and Morton Thiokol. The following table may help in keeping the players straight. Titles are from 1986 unless otherwise noted.

NASA (Name)	NASA (Title)	Morton Thiokol (Name)	Morton Thiokol (Title)
Aldrich, Arnold	Mgr., Nat'l. Space Transportation System Program Office, Johnson	Boisjoly, Roger	Applied Mechanics Engineer, Seal Investigation Team
Beggs, James	Former NASA Administrator	Brinton, Boyd	Manager, Space Booster Project
Crippen, Capt. Robert L	Investigation Liaison	Buchanan, Jack	Manager, Cape Operations
Hardy, George	Deputy Director of Science and Engineering, Marshall	Crippen, Capt. Robert L.	President, Aerospace Group (1996)
Houston, Cecil	Marshall's Resident Manager at the Cape	Ebeling, Robert	Engineering Department Manager
Lovingood, Dr. Judson	Deputy Shuttle Project Manager, Marshall	Garrison, Ed	President, Aerospace Operations
Lee, T. J.	Deputy Director, Marshall	Kilminster, Joseph C.	VP, Space Booster Program
Lucas, Dr.	Director, Marshall	Lund, Robert	VP of Engineering
Moore, Jesse	Assoc. Administrator for Manned Space Flight	McDonald, Alan	Director, SRM Project
Mulloy, Lawrence	SRB Project Manager., Marshall	Mason, Jerald	Senior VP, SRM Division
Reinartz, Stanley	Shuttle Project Manager, Marshall	Thompson, Arnie	Supervisor, Solid Rocket Cases
Wear, Lawrence	Manager, SRM Project office, Marshall	Wiggins, Calvin	VP and GM

CHAPTER III
NASA's Involvement—The Club
By Blattner

The place is the mission launch center at Cape Canaveral. The scene unfolding is recreated through the facts and testimony from the Presidential Commission to investigate the Challenger accident and interviews of personnel involved.

The time is 12:35 p.m. on January 27, 1986, and the talk around the launch center is about the "scrub" of the Challenger launch originally scheduled earlier that day. A launch manager takes a moment to call his wife to tell her he won't be home that night. "The decision to move the launch until the next day isn't technical. It concerns high winds around the launch pad and rough seas that might hinder a recovery of Challenger should there be an aborted flight. Hey, anyway, I'll be home tomorrow night," he adds.

But the launch manager's mild demeanor is deceiving. The pressure he and others feel has been building since the quick turnaround from the Discovery launch two weeks prior. The heavily delayed space Shuttle Columbia was supposed to have been launched the previous July. It took six months to fix numerous technical problems—six months of meetings and problem-solving sessions that have contributed to the fatigue some now feel.

Lawrence Wear (NASA, Marshall, Huntsville, AL), manager of the Solid Rocket Motor Project Office, spoke with Boyd Brinton, Thiokol's manager of the space booster project. Wear said a new temperature forecast predicted unseasonably cold temperatures overnight and at launch time. Wear asked, "Does Thiokol have any ofrficial concerns about the weather at Cape Canaveral?" Wear was referring to the fact that the coldest day for a shuttle launch previously was 53°F for STS 15; and that when that happened, there was evidence of rocket O-ring seal blowby or leakage.

Brinton said, "I don't know how much interest about weather there is, but I'll check and find out." He then phoned Arnie Thompson, Thiokol's supervisor of solid rocket's cases in Utah, and Alan McDonald, director of the solid rocket motor project, who is now at Cape Canaveral for the launch. McDonald calls Thiokol engineers in Utah and requests that they make an evaluation of how the weather at the Cape would affect the O-ring joints on the solid booster. "I'll phone you with weather temperatures all night long if I have to," McDonald says.

At 2:00 p.m. on January 27, at a NASA Mission Management Team meeting, there is no reference or direct conversation concerning how weather conditions might affect the O-rings. No one brings up Thiokol's concerns about the rocket seals in cold weather.

At 4 p.m., McDonald gets a call from Robert Ebeling, a Thiokol Engineering Department manager in Utah. Ebeling tells McDonald that he and others in engineering are very concerned about the weather at the Cape. "We just received word that the weatherman is projecting temperatures as low as 18 degrees some time by early morning," he says.

Ebeling says he has obtained launch-temperature projections from the operations center, and basically, the center feels it was going to be near freezing, or freezing, before midnight. "By 6 a.m. it could get as low as 22°F, and they're predicting a temperature of about 26° at the intended launch time," he tells McDonald. If this happens, it would be the coldest temperatures ever faced during an operational launch. Thiokol has no specific research on these temperatures but considers them as being outside the operational envelope of the solid rocket motors.

McDonald then calls Cecil Houston, NASA-Marshall resident manager at the Cape, and tells him of Thiokol's mounting concern about the weather and rocket motor's O-ring seals. Houston then relays the information to Dr. Judson Lovingood, NASA-Marshall Deputy Shuttle Project Manager. Houston asks Lovingood to set up a teleconference with senior project management personnel.

This includes George Hardy, NASA-Marshall Deputy Director of Science and Engineering, and various Morton Thiokol personnel.

Lovingood calls Stanley Reinartz, NASA-Marshall Shuttle Project Manager, and informs him of the planned telecom. It is now 5:45 in the evening, January 27.

The first telecom participants are Reinartz, Lovingood, Hardy, and numerous people at Kennedy, Marshall and Thiokol-Wasatch locations. Allan McDonald misses the meeting. But the conversation concerns the effects of low temperature on the O-rings. Thiokol presents its opinion that the launch should be delayed. There is a spirited discussion, but no decision is made about delaying the launch as the telecom ends.

Lovingood calls Arnold Aldrich and tells him about the telecom. He says it is possible that Thiokol will call for a launch delay. He says there is a possibility of a second telecom, and promises to keep Aldrich informed.

At 8:45 the evening of January 27, a second telecon begins. This time, Thiokol in Utah is represented by Jerald Mason, Senior Vice President of Thiokol's solid rocket division, Wasatch. Calvin Wiggins, a vice president and general manager at Wasatch, is there, as are other Wasatch personnel including Joe Kilminster, another vice president. Robert Lund, Vice President of Engineering, and Roger Boisjoly, a team member of the internal rocket seal investigation, is also represented. Arnie Thompson is also there.

NASA personnel include George Hardy, Deputy Director for Science and Engineering, Judson Lovingood, and Lawrence Mulloy, SRB Project Manager, Marshall.

The meeting begins with Thiokol presenting a history of the O-ring erosion and blowby—where smoke from the rocket motor actually escapes by the primary seal in the field joint. This presentation is detailed, including the results of subscale tests, previous flights, and static tests of solid rocket motors. Boisjoly says that the data indicated the timing function of the O-rings would be slower on the Challenger launch due to lower temperatures. "We don't have a high probability of sealing or seating. If the threshold is reached, then the secondary seal may not be capable of being pressurized, and that's the bottom line," he says.

The discussion moves on, and a Thiokol representative says, "The worst blowby of an O-ring occurred on the military launch of the space Shuttle Discovery (51-C). The temperature then was 53F, far warmer than the predicted temperature for the Challenger launch tomorrow." This remark gets this reply from Lund: "Challenger should not fly until the O-ring reaches at least 53°F."

Mulloy then asks Kilminster whether he can give assurances to launch. Kilminster replies that based on his opinion, and Thiokol's engineering recommendation, he cannot give such a recommendation. So, two high-ranking Thiokol personnel have said not to launch. Hardy enters the conversation and seems incensed by Thiokol's comments. McDonald and Boisjoly note this, writing down their observations. Hardy says, "I'm appalled by this decision not to launch." (Later, during Commission testimony, Hardy "spun" this statement by saying he was appalled by the data presented on joint-seal performance.)

Reinartz says, "I'm under the impression that the solid rocket motor is qualified from 40 to 90 degrees Fahrenheit." But the comment ignores the fact that the temperature prediction for the launch is predicted to be colder than 40°F.

Now Reinartz, Hardy, and other NASA personnel began to challenge Thiokol's conclusions and recommendations. Kilminster finally cuts off the discussion by saying that Thiokol needs a five-minute caucus to regroup. Although agreed to, the caucus takes more than thirty minutes. It is a tumultuous meeting with Thompson and Boisjoly voicing objections to launch. Lund also remains against a launch decision.

But then Mason says to Lund, "It's time to take off your engineering hat and put on your management hat." And then he says to Lund, "Remember that NASA is our largest customer." Finally Mason ends his comments with, "Just remember that the 'customer is always right.'" This is a clear urging of Lund to change his mind.

In Commission testimony Lund says, "I didn't realize until after that meeting, and after several days, that we had absolutely changed our position; but that evening, I guess I had never had those kinds of things come from the people at Marshall. We had to prove to them that we weren't ready, and so we got ourselves in the thought process that we were trying to find some way to prove to them it wouldn't work. We couldn't prove absolutely that the motor wouldn't work." Finally, in spite of some factual objections from lower ranking Thiokol engineers at the Cape and in Utah, Kilminster hammers out an agreement. Thiokol officially says in writing that there is a factor of three times the previous worst-case blowby; and even if the primary O-ring fails, the secondary will be in position—and will work.

McDonald seems confused and bewildered by this change of events. He decides—even facing possible dismissal—to try to stop the launch. He calls and tells his wife that he honestly thinks he will be fired for his objections to the NASA launch decision.

Now one hour before midnight, a second teleconference continues. The turnaround from Thiokol's objection to acceptance is complete. Kilminster reads the rationale for Thiokol's recommendation to launch. When he finishes, Hardy asks him to "put this recommendation in writing, and fax it not only to Cape Canaveral, but also to Marshall." As the meeting ends, the launch countdown moves forward.

At about 11:15 that night, McDonald decides he has to make a final plea not to launch. This is going against his own company's formal recommendation, but he can't turn back with so much at stake. McDonald gets Mulloy, Reinartz, and Jack Buchanan, Thiokol's manager of Cape Canaveral operations, on the phone together. In a prolonged conversation, McDonald challenges NASA to explain how it could justify a launch below the solid rocket motor's qualified temperature. The conversation goes on past midnight, but those at NASA seem committed to launch; they throw the responsibility back to Thiokol, saying the recommendation for launch is in writing from Thiokol. The NASA mindset is complete.

McDonald then tries a last-ditch effort. He is tired, alone, a single Thiokol employee speaking to the manager of his company's largest customer. But he has to say it. "I wouldn't want to explain to a Board of Inquiry why NASA launched Challenger, since there is a question about launching with O-rings at low temperatures, recovery ships heading into high seas, and icing conditions on the launch pad," he says to Mulloy.

"Don't worry about it. The launch it is not your concern," Mulloy replies—even though McDonald is part of the launch clearance process. McDonald, still on Mountain Time, has been involved in the countdown conversation for nearly 12 hours. Now after midnight, Canaveral time, he feels beyond exhaustion.

Mulloy then says, "Tell you what. Your concerns will be passed on as an advisory." Yet McDonald's concerns will not reach any managers in the higher chain of command.

At three in the morning of launch day, the Cape Canaveral ice crews inspect Launch Pad 39-B (Challengers' launch pad) and find large quantities of ice on the fixed service structure, mobile launch platform, and pad apron. They call it in.

By 5 that morning Mulloy, too, has been up almost 24 hours, and finally tells Dr. Lucas about McDonald's temperature concerns on the O-rings. Still, he shows Lucas a copy of the Thiokol fax containing the recommendation for launch. There is no further discussion of these concerns.

At about 7 in the morning and "counting," the flight crew is loaded aboard the Challenger for Mission 51-L. Commander Dick Scobee, Pilot Mike Smith, Mission Specialists Ellison Onizuka, Judith Resnik, Ron McNair, and Payload Specialists Christa McAuliffe and Bruce Jarvis, are all smiles and looking forward to their four days in space. They are now engaged in preliminary work as they have two periods of high activity scheduled, including a readiness check of a NASA Spartan satellite, once Challenger achieves orbit. The temperature of the Challenger's left-hand solid rocket motor registers about 25°F, and the crew reports patches of sheet ice on the lower segment and skirt of the left solid rocket motor.

At 8:00 a.m., NASA's Lovingood tells Lee, NASA-Marshall Deputy Director, about Thiokol's concerns and how the company had wanted to stop the launch; but after their caucus, they suddenly recommended launch. Then he says, "Thiokol has provided their recommendations in writing and the launch is a go."

An hour later the mission management team discusses the ice condition on the launch pad. But once again there is no mention of the cold weather or previously reported O-ring problems. Still, because of the weather and some minor problems, the launch is slipped to a later time. But by 10:30 the crew inspects the launch pad for a third and final time. Ice is removed from the water troughs. There are no further delays. And exactly at 11:38 a.m., Challenger lifts off to meet its destiny.

Macidull felt NASA management was acting like a secret society. NASA had turned into a club for the super elite, or the stratified culture of those NASA executives inside the manned spacecraft program. In this club you were either in or out. Outsiders, like most of the Commission staff, or U.S. taxpayers, had no membership. After all, they had not risked their lives on the forefront of technology for the good of all Americans. Of course, most "club" members had not done so either.

Even the thousands of dedicated NASA engineers were not members. They were only the worker drones. Only the elite management could get special perquisites and their own free forms of transportation. Certainly this was not unlike CEOs of many Fortune 500 corporations or even some political and religious groups in the nation and around the world. But the difference is that American citizens have always had a choice to invest or not to invest in a particular corporation or a religious organization. NASA management, through their political manipulations during the 1980s and the Shuttle program, not only focused on non-accountability for excesses but also put the safety of the worker drones—like the Challenger astronauts—on the back burner.

CHAPTER IV
The Roman Candle Maker
By Blattner

A major question at the time of the investigation—and one that is relevant today because of its ethical implications—remains: why didn't the manufacturer of the solid rocket booster, Morton Thiokol, stop the launch of Challenger if they were so certain that the rocket joints would be a problem in cold weather? Macidull says that part of the answer is obvious: "First, NASA's strategy was to force Thiokol to prove that the solid rocket motors were safe to fly. Or to put it another way, Thiokol had to prove that the rocket boosters would not work in cold weather. Next, Thiokol senior management was more concerned about the business decision of disagreeing with its largest customer, NASA, than about astronaut safety."

Many a corporate decision is made for financial, rather than quality or safety reasons. The dilemma of the launch decision for Thiokol was that it was going up against its largest customer.

Part of this answer can be found in an August 1986, *Aviation Week & Space Technology* news report stating that an investigation by the General Accounting Office (GAO) found Challenger's solid rocket motor had failed 2,075 possible quality control problems. In testimony before the House Government Operations Committee, the GAO said Thiokol's Shuttle booster facility in Utah had serious quality control deficiencies. Was the solid rocket manufacturer so used to anomalies that it began to overlook them?

Yet you didn't have to be a rocket scientist to understand that the O-rings were causing problems. There was STS-6, launched April 4, 1983, which had O-rings affected by the heat. STS 41B,

February 3, 1984, experienced O-ring erosion in both the right-hand nozzle joint and left solid rocket field joints. STS 41C, April 6, 1984, experienced erosion of the primary O-ring in the right-hand nozzle and heat damage on the left SRM in areas adjacent to blowholes in the putty of the secondary seal. To be specific, fifteen such O-ring incidents occurred in 24 launches, including Columbia, which had launched two weeks prior to Challenger.

That's more than half of the manned launches that went into space from the beginning of the manned Shuttle program from 1981 until 1986. Why didn't someone blow the whistle on the O-ring problem before seven people died?

Former NASA Administrator Jim Beggs says, "I never recall any pressure anytime to do a launch. When I was at NASA, my standing order was that we should never sacrifice flight safety for anything. I told everyone that I would stand by them if they felt in any way that the flight was jeopardized. And I know of no one who didn't feel the same way at the agency."

If this is true, then the NASA managers who literally coerced McDonald during those last hours before the launch didn't get the message. Still, Thiokol's guilt was not in its design of the solid rocket motor; it was in a corporate climate of pressure to keep its largest customer happy. How else could anyone rationally explain company managers—who were technically schooled in the problems of the solid rocket booster's field joints—changing a *no-go* to a *go* launch decision with so much engineering reason *not* to launch? The thing that drove those few brave people at Thiokol crazy was that their own senior management failed to listen to reason.

But some at Thiokol *did* try to stop the launch. From Commission testimony spread over pages 104 to 111 of the Presidential Report, we know that Alan McDonald tried to stop the launch.

Another young engineer at Thiokol, Robert Ebeling, was also in favor of stopping the launch. He called from Utah to Cape Canaveral, to warn of his grave concern that the launch was in jeopardy because of Canaveral's unseasonably cold temperatures. There were other lower ranking engineers such as Roger Boisjoly who did not want to launch in cold weather. And, according to

Commission testimony, even Robert Lund, the vice president of engineering, did not want to launch.

Therefore, a wide cadre of engineers at Thiokol recommended a no-go on that cold launch day in Florida. How Challenger was launched anyway is a story that begins inside of Thiokol.

During a Congressional hearing on the accident that took place after the Commission had adjourned, McDonald and Roger Boisjoly were asked why they didn't just pick up a phone and call the President of the United States to stop the launch. Says McDonald, "I thought that was the most stupid statement I had ever heard. Then Roger turns to me and says, 'You know, there was nothing preventing me from doing just that.'"

Boisjoly was outspoken about his Challenger experiences and vociferous in saying that Thiokol management had erred. When questioned by the Commission about the Thiokol decision to launch, he said, "I have been distressed by the things that have been appearing in the paper and things that have been said in general, and there was never one positive pro-launch statement ever made by anybody. There have been some feelings, since then, that folks have expressed that they would support the decision, but there was not one positive statement for launch *ever* made in that room."

Earlier, commissioners had pressed Boisjoly about what took place at Thiokol when the company caucused to change its engineering decision concerning launching Challenger. He said: "After Arnie [Thompson] and I had our last say, Mr. [Jerald] Mason said we have to make a management decision. He turned to Bob Lund and asked him to take off his engineering hat and put on his management hat. From this point on, management formulated the points for decision making regarding the launch. But there was never one comment in favor, as I have said, of launching by any engineer before or after the caucus."

We asked Boisjoly to comment about his Challenger experiences for this book. He refused, saying that he was "done talking about Challenger to reporters and others." Yet Boisjoly was evidently not finished writing about the accident. On an Internet Online Ethics homepage, he laid out his case for anyone who logged on. This is what he has written about the Thiokol change of heart concerning the launch decision:

“Some discussion had started involving only the managers when Arnie Thompson moved down the table to a position in front of the managers and once again, tried to explain our position by sketching the joint and discussing the problem with the seals at low temperature. Arnie stopped when he saw the unfriendly look in Mason’s eyes and also realized that no one was listening to him. I then grabbed the photographic evidence showing the hot gas blowby comparisons from previous flights and placed it on the table in view of the managers and, somewhat angered, admonished them to look at the photos and not ignore what they were telling us; namely, that low temperature indeed caused significantly more hot gas blowby to occur in the joints. I, too, received the same cold stares as Arnie, looks that seemed to say, ‘Go away and don’t bother us with the facts.’ No one in management wanted to discuss the facts; they just would not respond verbally to either Arnie or me. I felt totally helpless at that moment and that further argument was fruitless, so I, too, stopped pressing my case.”

What followed infuriated Boisjoly. He said the managers were struggling to make a list of data that would support a launch decision, yet the data actually supported a no-launch decision (see Appendix IV). During the closed manager’s discussion, Jerry Mason asked the other managers in a low voice, if he was the only one who wanted to fly, and Boisjoly said that no one answered him. “At the end of the discussion, the vote poll was taken by only the four senior executives present since the engineers were excluded from both the final discussion with management and the vote poll. The teleconference resumed and Joe Kilminster read the launch support rationale from a handwritten list, and recommended that the launch proceed as scheduled. NASA promptly accepted the launch recommendation without any discussion or any probing questions as they had done previously. NASA then asked for a signed copy of the launch rationale chart.”

Boisjoly says Thiokol management fully supported the original decision to not launch below 53°F prior to the caucus. “The caucus constituted the unethical decision-making forum resulting from intense customer intimidation. NASA placed Thiokol in the position of proving that it was not safe to fly instead of proving that it was safe to fly. Also, note that NASA immediately accepted the new decision to launch because it was consistent with their desires and please note that no probing questions were asked."

Another part of the reason that Thiokol did not stop the launch begins in 1943, when Thiokol Corporation management (Morton and Thiokol would merge in 1982) became curious about orders for liquid polysulfide (LP) polymer that began to arrive from Cal Tech's Jet Propulsion Laboratory. Thiokol soon discovered that LP made the best solid propellant fuel for rockets. Not long after, Thiokol's Aerospace Division was formed.

In the 1960s and 70s, the company was diversifying into numerous chemical businesses. Its solid rocket business was awarded lucrative contracts to build solid rocket motors for the Air Force's Minuteman missile and the Navy's Polaris missile. Then there were follow-on contracts for the U.S. Peacekeeper ICBM and the Navy's Poseidon missile. As the Apollo program came to a close in the early 1970s, NASA sought development of a manned, reusable space transportation system. A major consideration in this propulsion system was cost. In 1974, Thiokol won the $800 million contract to design and build the solid rocket motors to help launch the Space Shuttle. The first solid rocket-boosted Shuttle succeeded in 1981 with the flight of Columbia (STS-1).

Looking at the solid rocket boosters, one can't help but think that they look like overgrown Roman candles. Although technically highly sophisticated, they use the same principle of pyrotechnics perfected by the Chinese about 2,000 years ago. But from ignition to end of burn, a Thiokol recoverable solid rocket motor used to boost today's Shuttles has an average thrust of 2.6 million pounds and burns for approximately 123 seconds. Thiokol's rocket motors were almost always reliable from their earliest development.

While NASA declared the Shuttle operational after the fourth flight, it hid the fact that all was not well at the nation's largest solid rocket manufacturer. It was widely reported in the media that there were technical problems in producing and maintaining the reusable boosters, and not that Thiokol didn't have opportunities to structure the workforce for corrective action. A NASA document says that the Thiokol management style would not let anything compete or interfere with the production and shipping of boosters. The result, according to NASA, was a program that gave the appearance of being controlled, while it was actually collapsing from within due to excessive technical and manufacturing problems.

The Challenger accident itself fell heavily on the once-proud company. While disagreement among Thiokol engineers had always been a part of the environment, even part of a learning curve, there developed a personal loathing for those who disagreed, by those they disagreed with.

Shortly before the Challenger launch, the arguments between engineers widened into career-ending arguments. The Boisjoly statement tells what happened to change the company's attitude about launching Challenger. The Commission touched on this point; but the Commission did not know, much less, care about the degree of rancor that flowed through the halls of Thiokol's engineering department.

The split in the engineering division caused a major fracture in Thiokol's corporate culture. Some would lose control and later leave the company. Yet others would take the loss in stride, and some would continue to advance within the company.

Boisjoly was outspoken. He may have even set a type of record as being the most widely quoted engineer of that era. After the accident, his comments on the pages of publications such as *Aviation Week & Space Technologies* infuriated Charles Locke, then Thiokol's Chairman and CEO. They also served to drive a wedge between him and fellow engineers. The friction eventually led to Boisjoly leaving the company. His corporate demise was in plain view on the Internet.

After the accident, Boisjoly wrote about the split between senior management and the engineering staff. The first hint of this separation occurred during the preparation for the first closed-door testimony to the Presidential Commission. "I was given very little notice that a hearing would be held on February 14, 1986. I had approximately two hours total by myself prior to the hearing and was struggling to organize a set of notes to aid me during my testimony, while management had a publications department to prepare a formal set of professional viewgraphs for their version of the events leading up to the launch decision."

Meanwhile, at a pre-hearing meeting with management, the company attorneys advised those testifying to answer all questions with only "yes" or "no" and not to freely volunteer anything. "This advice was not followed by me nor Arnie Thompson and Alan McDonald, and there were obvious tense feelings between management and us after the testimony session," wrote Boisjoly.

Boisjoly's criticism wasn't only about Thiokol management. He saw MSFC (Marshall Space Flight Center) management manufacture conditions to cover up the real reason for the crash—the low temperature that contributed to the Challenger disaster. "I witnessed NASA spend a two-to-three-week side effort to prove that a Challenger joint close-out photo showed a twisted O-ring prior to mating. This was after I told them that the photo was only showing a track of smeared grease and not a twist because I had run a desktop test myself some time before to show that the O-ring could not twist as they were contending. I was proved right after their efforts failed to produce a twist in the O-ring, and they finally admitted that the photo showed a smeared grease track."

Previous to Boisjoly's testimony, he always prepared and presented his own material. Even his supervisors gave him total freedom because of their confidence. But after testifying for the Commission, all of this changed. Boisjoly said the president of the aerospace operations at Thiokol ordered that he should be kept isolated from NASA. "I didn't fully recognize the situation, and I continued to argue for full truthful disclosure while factions of Thiokol and MSFC management were fully content to tell only half-truths about the history of the development and production of the solid rocket boosters. I realized for sure in mid-April that I was actually being isolated from NASA and the main redesign activity, since mid-March while Thiokol management was telling me of my great importance to the redesign effort. I was being asked to furnish technical design information for the new designs, which was sometimes changed without my knowledge and was being presented to NASA by someone else with no copies of the final version of the presentation given to me for feedback. I was, in effect, actually isolated from the main redesign effort."

Boisjoly's fate was sealed. His working conditions kept deteriorating. Yet once again he was called to testify before the Presidential Commission in closed session on May 2, 1986. But the night before the meeting, Boisjoly, Bob Ebeling, and Alan McDonald met with the President of Aerospace Operations at Thiokol, Ed Garrison. Garrison opened the meeting with a few general remarks about the upcoming session with the Commission. But it didn't take long, according to Boisjoly, for him to be attacked for airing the company's "dirty laundry" in memos he had given to the Presidential Commission. "He also stated that Thiokol had suffered enough as a result of public disclosure but that we should continue to tell the truth, but we should consider the best way to state it before speaking. I quickly took exception to his remarks about me and said that I had simply tried to restore the truth in all testimony and I didn't consider my actions as airing dirty laundry. Bob Ebeling then spoke up with some support for me and he told Mr. Garrison that Thiokol should form a specific type of engineering development group. Garrison snapped at Bob and told him to shut up, and quit telling him how to run the company."

At the Commission session the next day, Boisjoly remembered that the conditions were as tense as he had ever seen them. "When Chairman Rogers asked McDonald and me about our current job assignments, we answered him and he was visibly upset because we were being punished for honesty in our testimony. The Commission then decided to release the closed-session testimony to the public and Thiokol received tremendous criticism from the Congress, the Presidential Commission and the news media."

That testimony turned out to be the end of Boisjoly's career at Thiokol, although he didn't know it at the time. He would return to Utah to face major morale problems within Thiokol since many of the engineers perceived his testimony as damaging to the company. Soon followed a meeting with the three top executives at Thiokol. That private meeting was held on May 16, 1986, with the CEO, the President of Aerospace Operations, and the Vice President of the Solid Rocket Motor Program.

"The meeting produced a very candid discussion of problems from our side but it was essentially one-sided with management telling us very little. The CEO even made it sound like we were on probation and if we worked hard and proved ourselves during the redesign activity, then everything would be forgiven. His attitude was certainly consistent with his criticism of McDonald and me in his previous statement to the *Wall Street Journal*. I believe it was at this meeting that the CEO made the statement that the company was doing just fine until McDonald and I testified about our job reassignments on May 2. He said that those statements caused the company more harm than all the previous releases," said Boisjoly.

Boisjoly had now been chastised and criticized before colleagues and ignored personally and professionally. Still he, McDonald, and Thompson, along with Thiokol management, were asked to testify at the House Committee on Science and Technology hearings held June 17-18, 1986. After these hearings, Boisjoly would finally break down.

"After my testimony to the House Committee, I could no longer endure the hostile environment at Thiokol so I took some time off at the recommendation of a company executive. During this period of being absent from my job, I realized that I could not subject myself again to the hostile environment, so I informed them that I would not return to work. I was then placed on extended sick leave from July 21, 1986, until January 18, 1987, at which time I qualified for long-term disability."

For two years after leaving Thiokol, Boisjoly would receive 60% of his former salary. But his disability compensation ended January 18, 1989. During this time, Boisjoly sued the company for $1.5 million, claiming mental anguish. He lost the case. Since then, he has been living in Mesquite, Nevada, serving as an expert witness for various Los Angeles law firms.

Alan McDonald supported some of what Boisjoly said in the ethics paper. But McDonald acknowledged that Boisjoly left Thiokol under his own volition and was not involved in any of the redesign of the O-ring on the new solid rocket reusable motor.

McDonald, Thiokol's manager of solid rocket motors during the time of the Challenger launch, was Boisjoly's boss during those years. While McDonald did take chances at speaking out during the last hours of the Challenger countdown, his engineering ability and focused approach to problem solving was far different than the thunder-and-lightning rhetoric of his subordinate.

McDonald stayed with Thiokol, and at this writing he is Deputy Director of Science and Engineering as well as Technical Director of Thiokol's propulsion department. But like all those we spoke to who were present at Cape Canaveral during the count-down, talking about the subject is painful. His comments were like a man talking about a life-threatening experience:

"We made a recommendation not to launch; but later that was changed and we said it was okay to launch, even though temperatures were colder than at any time in prior launches. To be honest about it, I was completely confused about that decision and couldn't understand how we supported that recommendation. Yes, there were some data that were inconsistent, but enough evidence showed we had a problem and the temperature was going well below where we had that problem. You don't go launching Shuttles like you're playing Russian roulette."

While McDonald does not deny that the O-rings prior to the Challenger launch were problematic, he does dispute NASA's role. He said flatly that NASA had a chance to fix the problem months before the Challenger accident. "In August, 1985, I made a full presentation to NASA Headquarters about the pressure seals on the solid rocket motor. We said at that time it was a very significant issue, something that had to be fixed. The reason we did this was because in April of that year we had a problem with a seal on a rocket motor that we had never seen before. It wasn't in the field joint; it was in the nozzle case joints. So we wrote it up as a problem with a primary O-ring seal and part of the secondary. During the discussion at NASA, we included information and said NASA should shut down the fleet until the problem was fixed. We said, 'yes, you ought to do that until we understand why this happened, and whether it will happen again.'"

McDonald says Thiokol went through all the drawings, procedures, and designs of that particular O-ring with NASA. "In that review we learned that when we were pressurizing between these two seals, to check and make sure they were operating properly. We couldn't verify whether the seals were seating properly because the pressures we were using to run that test were too low. You couldn't even tell if there was an O-ring. All the subsequent solid rocket motor O-rings were checked with a much higher pressure leak check to make certain they weren't leaking."

As a result, Thiokol made a recommendation to NASA in 1985 to restrict the motion in the joint. "The problem was not with the O-ring, it was with the motion in the joint. The O-ring had to maintain contact with the metal surfaces when it was moving. If you don't move the metal surfaces you don't have to worry about temperatures or problems because you're not asking the O-ring to hang onto a moving surface and expand into the surface opening. That was the problem with O-rings. They got cold and had no resiliency to expand when the surface was open. But that recommendation was not accepted by NASA," McDonald said.

Instead, NASA chose to ignore Thiokol's major finding. "NASA released a memorandum that said that there were thousands of these O-rings, and now there had been a problem. The implication was that Thiokol wanted more money to change the rocket motor cases and put in a new feature, so therefore NASA found it cost prohibitive," McDonald says.

During the final few hours of the Challenger countdown, NASA managers Mulloy and Reinartz would badger McDonald for "hard proof" that the solid rocket motor O-rings would fail in cold weather, ignoring the information that had already been presented. "Clearly, there was no one in Thiokol systems engineering who could prove the O-rings would fail. That's a true statement. We did not have engineering tests that would prove that the rings would fail. NASA was pushing people to say 'well, if they won't fail, then it's safe to fly.' That's a whole different logic train. NASA forced us to make a management decision, not a safety decision, and Lund had to put on that 'management hat' and change his own personal opinion—that it was a mistake to launch."

The Commission report, in part, states:

"NASA's system for tracking anomalies for Flight Readiness Reviews failed, in that, despite a history of persistent O-ring erosion and blowby, flight was still permitted. It failed again in the strange sequence of six consecutive launch constraint waivers prior to 51-L [Challenger] permitting it to fly without any record of a waiver, or even an explicit constraint. Tracking only anomalies those were outside the data base of prior flight allowed major problems to be removed from and lost by, the reporting system."

The Commission concludes its findings on the above dialogue by saying:

> "A careful analysis of the flight history of O-ring damage and low temperature ... 'showed' ... neither NASA nor Thiokol carried out such an analysis; consequently they were unprepared to properly evaluate the risks of launching the 51L mission in conditions more extreme than they had encountered before."

These comments are accurate as far as they go. The accident, however, occurred because of NASA managers' intransigent view that the flight had to go.

Speaking in their defense, Arnold Aldrich, NASA's former Manager, National Space Transportation System Program Office (Johnson Spaceflight Center) says, "I can only conclude what has been reported, and that is that the people responsible for that work in the solid rocket booster project at Marshall [i.e., Hardy, Reinartz and Mulloy] believed that the concern was not of a significance that would be required to be brought forward because clearly the program requirements specify that critical problems should be brought forward to Level II and not only to Level II but through myself to Level I."

The Shuttle Flight Readiness Review used in the 51L mission consisted of a step-by-step process established by NASA program directive, SPO-PD 710.5A, which was created on September 26, 1983. The process, simply stated, certified the readiness of all components of the Space Shuttle "stack" at the time of launch. A structured Mission Management Team meeting—L1—was held 24 hours prior to each scheduled launch. The agenda requires managers to close out any open work or any Flight Readiness Review action item. It was a standing practice of Level I and Level II officials to encourage the reporting of new problems or concerns that might develop in the interval between the flight readiness review, the L1 meeting, and between the L1 and the launch.

After hearing the testimony of NASA and Thiokol managers, the Commission concluded that failure in communication at the launch center resulted in the decision to launch Challenger. " ... based on incomplete and sometimes misleading information, a conflict between engineering data and management judgments, and a NASA management structure that permitted internal flight safety problems to bypass key Shuttle managers." But the Commission did not specify who precisely was responsible for the decision to launch Challenger. There was no accountability for those who caused the accident by pushing to launch Mission 51L.

McDonald thought some of the problems with the decision-making process came from exhaustion. He said that the delay on that launch was the longest encountered in any of the previous 24 Shuttle missions. "The Columbia launch two weeks in front of Challenger was delayed six or seven times, and it was tearing NASA apart. In a couple of years the NASA manifest showed that they were going to launch 24 times a year, and they couldn't even get this bird off the ground. It was also threatening to delay Challenger, their bread-and-butter launch. After the mission, Challenger was supposed to turn around and be launched again from Vandenberg Air Force Base in California on a military mission. It was a schedule that few of us could believe."

Still, how could Thiokol change its decision concerning the O-ring and say it was okay to launch? McDonald answered by asking, "Who was Thiokol's largest customer? NASA. They didn't agree with our recommendations. Their managers didn't agree that we couldn't launch. NASA had announced one week prior to the launch that they were going to seek a second source for the solid rocket boosters. Here you had a dissatisfied customer who was the company's largest, and unless we were convinced that we had made a mistake—and I reemphasize that we had no hard proof that the O-ring would fail—we were vulnerable to NASA. I always personally thought it was in the NASA manifest that 'if you don't know, you don't go.' We had some conflicting data with the seals. Everybody knows that we make tactical ICBMs that work at *minus* 60°F, but they didn't have opening joints like the Challenger's solid rocket boosters."

Thiokol had statically tested Challenger-type solid rocket motors in the 40°F range in Utah. The problem was that on a static rack, the field joints could be artificially patched up. "We viewed the assembly as an artifact of a horizontal assembly. The crew saw those holes in the joint and plugged them up while the booster was in the test stand. Well, we saw the same size holes in the booster recovered from the ocean."

McDonald thinks the failure leading to the Challenger accident boils down to NASA putting too much pressure on its crews to launch in fast turnaround. "Those managers could analyze safety and they could analyze failure, but they couldn't analyze themselves in failure. So Challenger must have been safe, right? Well, five of the six solid rocket motor joints worked; so they were 87 percent correct. Using that logic we were 13 percent right."

Nevertheless, by the middle of April in 1987, things would turn around for Thiokol. The engineering staff would bury problems and work on the fixes for the problem O-rings. McDonald said it was an upbeat time. "We had something to do, to fix a problem rather than simply talk about one."

The Challenger accident halted production of the company's solid rocket motors from 1986 to 1988. Thiokol's engineering staff devised what would become multiple design improvements of the reusable solid rocket motor (RSRM). In this design, there would be no troublesome O-rings. In 1992, Thiokol successfully test fired its Castor 120 motor, a low-cost, 120,000-pound solid rocket motor, the company says was designed for the next generation of expendable launch vehicle solid propulsion motors.

By 1996, Thiokol had created the Thiokol Aerospace Group and selected veteran Astronaut Robert L. Crippen as its president. Is there a connection between Crippen's days at Cape Canaveral after the accident and his appointment? On October 18, 1996, James R. Wilson, Thiokol's CEO, said of Crippen, aged 59 at the time, "His tenure with NASA spanned 26 years including four Space Shuttle flights. He served 30 years with the U.S. Navy as an aviator."

Obviously it is not often a person can start their career at three years old. But Crippen did not serve in the U.S. Navy as an aviator for thirty years. This is the type of hype that is neither honest nor productive. It simply gives the impression, to anyone hearing these inaccuracies, that the speaker regards his fellow citizens as stupid. And there may be some truth in this regard. It would have been accurate for Mr. Wilson's endorsement to say, *He was a designated naval aviator for thirty years on loan to NASA for twenty-seven years, and could have been called up for active duty in the*

Navy at any time to serve his country in a capacity consistent with a national emergency.

This point is made because, throughout the era of Challenger, there were many glib statements and references made that were untrue. Paraphrased examples: "This system is now operational," result – remove the ejection seats and parachutes. "This O-ring problem is a nuisance," result – change the management rules and ignore the problem. "The temperature is a problem for this Challenger launch. We, the NASA management, do not like this problem," result – cover your tracks and ignore it. And, probably the biggest insult, and which is not paraphrased: "It's about time you took off your engineering hat and put on your management hat." Management was not going into space. Management did not have their physical lives on the line. Management as a body could not be called engineers. Did "management" actually think that it could control the laws of Mother Nature? Evidently that *is* what they thought at the time.

CHAPTER V
The Disconnect
By Macidull and Blattner

Fatigue as a Contributing Factor

More than 5,000 people worked at Cape Canaveral in processing the Challenger in January 1986. These people were responsible for assembly, fueling, launch pad preparation, and other pre-flight requirements. Investigators for the Commission found that some workers had been on duty more than 80 hours the previous week, resulting in exhausted launch managers. Said the Commission, "The decision to launch should have been based on engineering judgments. However, other factors may have impeded or prevented effective communication."

Sleep deprivation can do strange things to the human body. Pilots tell of sleeping at the controls—with their eyes wide open. Truck drivers have been slow to react, driving right into an accident. To such a condition, add the lack of proper food, an excessive amount of coffee, tea, or soft drinks with large doses of caffeine consumed over long periods of time. Then stir into the mix the pressure caused by launch scrubs and fast turnaround times, and you have the recipe for major human malfunction or, more accurately, a human factors problem.

Such a problem faced the Challenger launch crew. McDonald and the other Thiokol personnel came in for the launch with body clocks still on Mountain Time. While people such as Hardy and Reinartz were in the Eastern Time zone, they had been under stress and working long hours not only because of the Challenger scrubs and reschedules but also because of the Columbia launch two weeks prior. Still, on January 28, 1986, NASA's premier launch team stood ready to commence its ninth Shuttle mission in a little over a year.

Thiokol Disconnects (by Macidull)

My first indication of a management disconnect was early in the Commission investigation. My research uncovered a telefax sent on the morning of the accident from Morton Thiokol, Wasatch Division, to NASA at both Cape Canaveral and the Marshall Space Center. The title of the telefax was "MCI Assessment of Temperature Concern on SRM (51L) Launch." I insisted that it be put in the report … Page 97, Volume I. I was amazed; here was a message in the form of points of information and decisions. The first point said that the weather was 20°F colder than on any other Shuttle launch. The second point said the temperature data was not conclusive on predicting primary O-ring blowby. Then there were several other points of engineering jargon, followed by the last main point, which said "MCI recommends STS-51L launch proceed on 28 January, 1986."

It is amazing that Thiokol, after first stating that it didn't know what would happen based on all available data, recommended launching anyway. Talk about a disconnect! Then to add insult to death, the two people best qualified to make this decision concerning their own physical well being and to fulfill their responsibilities for their crew, Scobee and Smith, were kept completely out of the decision-making process. After all, they were only pilots and engineers whose lives were on the line. They were not "managers." It is important that this disconnect be noted. It should go down in history as an example of technology getting so far ahead of management that block-diagram, pecking-order mentality started killing people.

Prior to the accident, neither NASA nor Thiokol management understood the mechanisms by which the joint sealing action took place. Thiokol was a premier defense contractor responsible for solid rocket missile motors, and yet their top management didn't know how a field joint worked. NASA was responsible for launching a $1.5 billion spacecraft with a human crew. They, too, had no understanding of how a field joint worked in relation to temperature.

Not only were NASA and Thiokol managers ignorant about how the joint sealing action took place—they were also poorly equipped to listen to their engineers. Both sets of managers were ignorant by education, background, and experience. Even if there had been a mislabeling of the criticality of the joints, these managers knew there was an O-ring problem with past launches when the solid rocket motors had dangerous erosion and blowby.

As McDonald indicated, Thiokol had been to Washington months before Challenger to tell NASA that there was a problem and to beg the agency to change the design. However, the Commission was riddled with vague and noncommittal language when it came to the responsibility and accountability for who should be charged in the negligence leading up to the Challenger launch.

The Commission report, page 176, states, "The capabilities of the system were stretched to the limit to support the flight rate in the winter of 1985/86. Projections into the spring and summer of 1986 showed a clear trend; the system, as it existed, would have been unable to deliver crew training software for scheduled flights by the designated dates. The result would have been an unacceptable compression of the time available for the crews to accomplish their training."

This FAX says two basic things: (1) We (managers) do not know what will happen under these new and untried environmental conditions, and (2) Launch anyway with people aboard. This FAX is also an example of technology getting so far ahead of management that block-diagram, pecking-order mentality started killing people.

JOE C. KILMINSTER'S FAX TO NASA THE NIGHT BEFORE CHALLENGER'S LAUNCH - ONE PAGE DATED 9:45 PM MOUNTAIN STANDARD TIME

27 JANUARY 1986

MTI ASSESSMENT OF TEMPERATURE CONCERN ON SRM-25 (51L) LAUNCH

- 0 CALCULATIONS SHOW THAT SRM-25 O-RINGS WILL BE 20° COLDER THAN SRM-15 O-RINGS
- 0 TEMPERATURE DATA NOT CONCLUSIVE ON PREDICTING PRIMARY O-RING BLOW-BY
- 0 ENGINEERING ASSESSMENT IS THAT:
  - 0 COLDER O-RINGS WILL HAVE INCREASED EFFECTIVE DUROMETER ("HARDER")
  - 0 "HARDER" O-RINGS WILL TAKE LONGER TO "SEAT"
    - 0 MORE GAS MAY PASS PRIMARY O-RING BEFORE THE PRIMARY SEAL SEATS (RELATIVE TO SRM-15)
      - 0 DEMONSTRATED SEALING THRESHOLD IS 3 TIMES GREATER THAN 0.038" EROSION EXPERIENCED ON SRM-15
  - 0 IF THE PRIMARY SEAL DOES NOT SEAT, THE SECONDARY SEAL WILL SEAT
    - 0 PRESSURE WILL GET TO SECONDARY SEAL BEFORE THE METAL PARTS ROTATE
      - 0 O-RING PRESSURE LEAK CHECK PLACES SECONDARY SEAL IN OUTBOARD POSITION WHICH MINIMIZES SEALING TIME
- 0 MTI RECOMMENDS STS-51L LAUNCH PROCEED ON 28 JANUARY 1986
  - 0 SRM-25 WILL NOT BE SIGNIFICANTLY DIFFERENT FROM SRM-15

Joe C. Kilminster

JOE C. KILMINSTER, VICE PRESIDENT
SPACE BOOSTER PROGRAMS

FAXED TO:

MSFC # 205-453-5725

KSC # 305-867-7103

9:45 PM MST

27 JAN 1986

MORTON THIOKOL, INC.
Wasatch Division

INFORMATION ON THIS PAGE WAS PREPARED TO SUPPORT AN ORAL PRESENTATION AND CANNOT BE CONSIDERED COMPLETE WITHOUT THE ORAL DISCUSSION

Listening Impairment and Lack of Knowledge

As McDonald indicated, Thiokol had been to Washington months before the Challenger launching to tell NASA that there was a problem and to beg the agency to change the design. However, the Commission was riddled with vague and noncommittal language when it came to the responsibility and accountability for who should be charged in the negligence leading up to the Challenger launch.

Arguments and Prophecies (by Blattner)

At the eleventh hour, McDonald argued with Reinartz and Mulloy about the qualification and launch commit criteria. Arguments were not unusual, but the tone and intensity were, especially since McDonald was Thiokol, and Reinartz and Mulloy were NASA, Thiokol's largest customer. Reinartz stated that the solid rocket motor was supposedly qualified from 40°F to 90°F. During Commission testimony, McDonald said:

"... and Mr. Mulloy said well, 40 degrees is propellant mean-bulk temperature, and we're well within that. That is a requirement. We're at 55 degrees for that, and the other elements can be below that as long as we don't fall out of the propellant mean bulk temperature. I told him I thought that was asininc because you could expose that large solid rocket motor to extremely low temperatures—I don't care if it's 100 degrees below zero for several hours—with that massive amount of propellant, which is a great insulator, and not change that propellant mean bulk temperature by only a few degrees, and I don't think the spec [specification] really meant that. I knew everything [within the rocket motor] was qualified to that temperature, in my opinion. But we were trying to qualify that case itself at 40° to 90° for the filament wound case."

McDonald, Reinartz, Mulloy, and a seemingly large cast from NASA would go at it into the wee hours of the launch morning. Finally, in frustration, McDonald said, "I wouldn't want to stand up in front of a board of inquiry and say that I went ahead and told them to go ahead and fly this thing outside of what the motor was qualified to do."

Such a comment may not sound heated if someone says it at a cocktail party, but at the eleventh hour before seven people were sent to their deaths, such a prophecy sends chills.

NASA Disconnects

But the launch management team didn't know about any of the dialogue among McDonald, Hardy, Mulloy and Reinartz. At a mission team meeting on January 25, no technical issues came up at the meeting or in the documentation, and all Flight Readiness Review actions were reportedly closed out. Mulloy subsequently testified concerning the flight readiness review record about O-ring concerns:

Chairman Rogers:	*"Why wasn't that a cause for concern on the part of the whole NASA organization?"*
Mr. Mulloy:	*"It was cause for concern, sir."*
Chairman Rogers:	*"Who did you tell about this?"*
Mr. Mulloy:	*"Everyone, sir."*
Chairman Rogers:	*"And they all knew about it at the time of 51L?"*
Mr. Mulloy:	"*Yes, sir. You will find it in the Flight Readiness Review record that went all the way to the L [level] 1 review."*

Following is an exchange between Chairman Rogers asking four key NASA officials—Smith, Canaveral Space Center Director; Thomas, Launch Director; Aldrich, Shuttle Program Director; and Moore, Associate Administrator for Space Flight about their knowledge of the Challenger launch.

Chairman Rogers: *"By way of a question, could I ask, did any of you gentlemen, prior to the launch, know about of the objections of Thiokol to the launch?"*

Mr. Smith: *"I did not."*

Mr. Thomas:	*"No, sir."*
Mr. Aldrich:	*"I did not."*
Mr. Moore:	*"I did not."*
Commissioner Hotz:	*"Mr. Thomas, you are familiar with the testimony that this Commission has taken in the last several days on the relationship of temperature to the seals in the solid rocket booster?"*

Mr. Moore:	*"Yes, sir, I have been here all week."*
Commissioner Hotz:	*"Is this the type of information that you feel that you should have as launch director to make a launch decision?"*
Mr. Thomas:	*"If you refer to the fact that the temperature according to the Launch Commit Criteria [LCC] should have been 53 degrees, as has been testified, rather than 31 [degrees], yes, I expect that to be in the LCC. That is a controlling document that we use in most cases to make a decision for launch."*
Commissioner Hotz:	*"But you are not really very happy about not having had this information before the launch?"*
Mr. Thomas:	*"No, sir. I can assure you that if we had had that information, we wouldn't have launched if it hadn't been 53 degrees."*

Said the Presidential report: *"It is disturbing to the Commission that, contrary to the testimony of the solid rocket booster project manager [Mulloy], the seriousness of concern was not conveyed in the Flight Readiness Review to Level I and the 51L readiness review was silent."*

So Mulloy misrepresented fact. It is a criminal offense to misrepresent facts to a presidential commission. Mulloy was never charged.

The Criticality Issue (by Macidull)

A part marked "Criticality 1" in the solid rocket motor simply meant that the part was critical to the booster's performance and did not have a backup system. Therefore, if there had been any question as to whether the part would work, or if there had been any type of anomaly with the part whatsoever, the booster could not fly until the problem was fixed. Conversely, if a booster part was marked as "Criticality 1R," it meant that the part had some form of backup. The "R" stood for redundant. Even if the original part failed, the reasoning went, the backup would take over. Therefore, parts with a Criticality 1R rating could be flown, even if there was an anomaly or problem with a previous flight. The O-ring seals were not on the Criticality List. They were part of a criticality system. The O-ring itself was not listed as a criticality part, and it was not 1R as stated by Mulloy since no such rating was ever applied on the official Criticality List. Yet Mulloy stated specifically that the seals were 1R. Furthermore, the dialogue indicates that the commissioners were buying into this wrong assumption.

A complicating factor was that engineers at both NASA and Thiokol had established the 1 and lR ratings. That may not seem like a problem; but it was, because they each had developed their own version of the list, and they ignored the 1982 Criticality List that was their regulatory requirement concerning critical parts of the rocket booster. North American Rockwell, during the building of the first space shuttle, created the official Criticality list. Early in the space Shuttle program, NASA agreed for safety reasons that this list should be adhered to. So, Thiokol applied to North American to have the solid rocket motor components put on this list. The list stated that the solid rocket booster field joint was Criticality 1. But there was no listing for the seals, as they were part of the joint.

The core of the problem was that the field joints moved into different positions during flight and the O-rings therein were subjected to the effects of different temperatures and dynamic pressures. NASA basically said that, though the field joint was single and critical, it was partly comprised of two parts that looked the same; therefore, they simply designated it as a redundant system without any research or safety considerations.

But it wasn't only the launch managers who were confused. NASA's top manager at the time, Associate Administrator for Manned Space Flight, Jesse Moore, had not impressed anyone on the Commission with his knowledge about what took place before the accident. Moore resigned shortly after the accident, but records show that he never bothered to attend any of the Flight Readiness Review meetings before any flight—*not any* of these crucial meetings before twenty-five launches during a five-year period. NASA's top manager of the manned spacecraft flight program apparently couldn't be bothered with flight concerns.

The criticality issue is important because it was central to the safety of the solid rocket motor. It is an issue that confused not only NASA management but also the commission itself. This can be explained by an exchange between commissioners and NASA's Lawrence Mulloy:

Chairman Rogers: *"It seems to me that the question and answer has become extremely important. Let's be sure that the record is clear on that. I thought Mr. Armstrong's question went right to the heart of the matter, and I'm not sure that I understood the answer. Is it possible to read that back?"*

Commission Reporter: *"Specifically with regard to the joints on 51-L [Challenger], based upon your analysis of the measurements that you have on those joints, would you expect that after full flight pressurization of the motor you would have a secondary seal? Or, to put it another way, do you think on these joints you had a Criticality I or a Criticality IR?"*

Chairman Rogers: *"Could we get an answer just to that one? Was that Criticality 1 or Criticality 1R?"*

Mr. Mulloy: *"Yes, sir. It was our judgment that we were redundant at ignition. As Dr. Ride has pointed out, there was a timing function. We did not conclude on that night [January 27, 1986] that the primary seal would not function and seal. That was inconclusive.*

If under the conditions described, if the primary seal did function and seal, then we have joint rotation, then the primary seal subsequently fails, I believe the question was that, under the conditions of temperature, would you anticipate that you had a worst-case condition wherein the secondary seal would not be in a position to energize.

I said that might be the case. It has not been shown what we had was resiliency data that says that the metal under rotation lifts off the seal. It has not been shown that that seal would not function under that condition."

Commissioner Hotz: *"Mr. Mulloy, in the interest of clarity, could you just simply state, was it 1 or IR?"*

Mr. Mulloy: *"It was 1R, sir."*

Blowby (by Blattner)

There was a second and equally crucial part to the criticality problem: Columbia mission 51C was launched two weeks before the Challenger. Columbia's solid rocket motors had the worst condition of blowby of any of the other twenty-three launches. Columbia's left solid rocket motor field joint had an 80-degree arc of black grease between two of its O-rings, and the right motor had a 110-degree arc of black grease between two of its O-rings. These were telltale signs of blowby. Thiokol engineers said this was because of O-ring squeeze due to the low temperature at launch time and the O-rings not filling up the holes in the field joints. Such blowby indicated that the propellant was escaping from the motor case, and under similar conditions there could well be a fire or worse. Said Thiokol's Boisjoly in testimony, "Again I brought up that the solid rocket motor 15 [flight 51C] had a 110-degree arc of black grease, while our solid rocket 22 [flight 61–an earlier Challenger mission, October 1985] had a relatively different [smaller] amount which was less and wasn't quite as black. I also stopped when it was apparent that I couldn't get anybody to listen."

So if Columbia had had this problem, and it happened right before Challenger, and it involved the solid rocket field joints, then *why,* under the terms of the Criticality List, weren't the field joints fixed before the next launch? This list, originally created on December 17, 1982, warned that a failure to correctly identify and fix Criticality1 field joint problems could result in: " ... loss of mission, vehicle, and crew due to metal erosion, burn-through, and probable case burst resulting in fire and deflagration."

Did the mislabeling and use of those criticality items in the solid rocket motor spell disaster for Challenger? From all the commission testimony, it seems that the NASA managers were not listening to Thiokol, even with much doubt about the seals.

National Security Disconnects (by Macidull)

The national security of the U.S. was severely compromised as a result of the Challenger explosion, leaving the U.S. vulnerable to advances by other countries for a period of nearly 3 years. Appendix III (by Blattner) includes a brief history of NASA and its symbiotic relationship with the DOD that led to this vulnerable position.

In brief, the satellite launch capability of the U.S. was compromised by the agreement between NASA and the U.S. Air Force to shut down the missile line and to restrict all satellite capability weight-wise and dimensionally to a single source—the shuttle system. When the Shuttle accident occurred, the U.S. missile capability, both military and domestic, ceased.

After nearly four decades, billions of dollars, and human sweat and sacrifice, America's manned space program wasn't some cavalier manifestation of an ill-conceived procedure. It had evolved into a carefully established program based on logic, scientific fact and personal experience requiring it to always err on the side of safety. In the past, if there was a known problem, it was fixed or the mission didn't go.

The Challenger explosion was an anomaly in a program constructed of building blocks of previous efforts—Mercury, Gemini, and Apollo—all of the manned space programs up until Challenger. Such building-block rationale assumes that logical thinking and procedures would be used concerning the criticality criteria in the Shuttle program. Still, what good is all that evolution if launch managers disconnect from the system and don't use proven procedures?

Such restrictions, including those on the field joints, were formalized as orders and distributed throughout both NASA and Thiokol management and subcontractor management. However, both NASA and Thiokol top management simply, informally, and on their own, ignored the established safety considerations to which they were required to adhere. It was a terrible example of ignorance for men who were billed as the best that formal education and on-the-job experience could create. It left in shambles the hard lessons developed by decades of NASA engineers and through the physical risks of many astronauts. Those involved in this final countdown decision clearly did not understand the relationship between temperature and O-ring problems.

CHAPTER VI
NASA Cover-Ups And Obfuscations
By Macidull

E-202 Film

I first encountered what can only be called a cover-up during an early trip to Cape Canaveral. My first commission trip to the Cape was in February 1986 to review the photographic evidence of the Challenger explosion that NASA had gathered. The head of the NASA photographic department, Mr. Charlie Stevenson, gave a briefing to the commissioners and staff in a room about an eighth the size of a basketball court. It was a dog-and-pony show, with 8 x 10 launch photos arranged on the walls. NASA had several movie cameras filming the Challenger, yet this film was missing from the presentation.

I stood in the back of the room during this presentation and noted a light table in the corner with a NASA technician next to it. I walked over and noted a reel of 70mm film set up on a spool. I asked the technician if it was of the Challenger. The technician said it was. Then I asked the technician to run it. As he cranked the handle, I saw beautiful and detailed positive color celluloid pictures of the Challenger rising into the sky and asked him to run it again and again.

The commissioners drifted over to the light table, and Stevenson came over to explain it. It was the Challenger launch. When I asked why there were only 58 seconds of the flight on the reel, he said it was because camera number E-202 had failed.

I taped the dialogue on a small recorder. It had a red sound-activated light, and it stuck out of the top of my left breast suit jacket pocket in plain view of everyone who cared to see. To me it was an end-of-story explanation, a why-do-you-ask attitude. Stevenson, to me, seemed irritated that this moving E-202 film was discovered and explained that there was no different information on it that wasn't also part of his planned show.

After everyone left, I went back to the light table and inspected the end of the film. There was no fading of light or other aberrations in the last few frames that could indicate a failing or failed camera. There were no torn sprocket holes. Then I looked closely at the last frame and ran my thumb over the end. It was a clean break. I held the edge up to the light. It was slightly curved. A clean break slightly curved? The film had been cut with a pair of scissors.

A month later, the missing 70mm film from camera E-202 still bothered me. I asked Commissioners Covert and Hotz to fly down to the Cape and go through NASA's photo files since the Commission needed more pictures for the report.

The next day at the Cape we started going through the files, laying photos out on a table. While Covert and Hotz were busy, I decided to look around. In another room was a large mainframe computer. Next to it were two technicians and a light table hooked to the computer. I introduced myself and looked at the 70mm film of an explosion as it lay on the light table. I asked if it was the Challenger explosion. They said yes. I asked, "From camera E-202?" They said yes and were very cooperative. During the conversation, they said that it had taken a few weeks to get the $25,000 circuit board that enabled digitization of the images (which, of course, allows video altering). Their task was to enable computer enhancement of the film and to put it on videotape for other types of analyses. This film had the complete E-202 footage through the explosion and for many seconds afterward.

Since I wasn't a commissioner, but now had good reason not to trust NASA management, I went to get Covert and Hotz. They looked at the detailed, high-resolution color film of the explosion and aftermath in amazement. Had I been the only outsider to see it, NASA easily could have again denied its existence.

The commissioners and I flew back to Washington that afternoon. I went straight to Astronaut Brewster Shaw, on the staff, and said that he should go down to the Cape and view the E-202 film, that the crew capsule came out of the fireball intact and not been rotating to any incapacitating degree, maybe one or two RPMs at most. I emphasized that Brewster's interpretation, as an engineer, pilot, and astronaut, would be helpful. Brewster looked shocked and said something like, "That's not at all what I wanted to hear." I said that he was part of this investigation, and that what he may personally want to hear or see was not germane to the effort.

The next morning a photo of the explosion from that light table appeared on the front page of the *Washington Post*. That evening at a restaurant with a few staff members, Brewster took me aside and, referring to the photo in the Post that morning accused me of "leaking" the photo.

Shaw mentioned that we were all, including the commissioners, given written instructions by Rogers not to talk to the media. I responded that he, Brewster, didn't seem to understand his own organization, and in my estimation once the film had been discovered by the Commission, especially so late in the game, NASA immediately released copies to the media to escape the accusation of a cover-up. After this confrontation, Shaw and I talked little for the rest of the Commission tenure.

During the final Challenger radio transmission, pieces from the external tank explosion ripped some of the Shuttle skin. At nearly twice the speed of sound (Mach 1.92) at 46,000 feet, the air pressure against a non-smooth or non-streamlined object would cause havoc. In Challenger's case, the thin air at that speed tore the ET and the Shuttle apart. Pieces of wing and engines clearly could be seen coming out of the cloud on the E-202 film. The outer shell of the crew capsule was ripped off. But the crew capsule itself was built to withstand tremendous force and was strong enough to remain pressurized in space. It came out of the cloud intact, trailing an umbilical of wire bundles that had been connected to the cargo bay and other aft components, including the Main Engines (MEs) in the tail of the orbiter. And it, the crew capsule, wasn't rotating much.

This is the point in the flight where nearly all information from NASA headquarters ended. The Presidential Commission on threat of a court order later gleaned some additional information. The rest of the flight history, along with what the commission obtained after the explosion, was based on radar information, voice recordings, photos, and wreckage—information that was both recovered and "released" under NASA's supervision.

Statistical Games

Early on in the Shuttle program, NASA began manipulating numbers. There was no reliability study done for the Shuttle as a system. However, the White House Office of Science and Technology had a requirement to provide the President with failure statistics of any and all nuclear components that the U.S. planned to send into space.

In the cargo bay of this last Challenger flight was a nuclear powered Gallileo satellite. It was called an RTG powered satellite, containing radioisotope thermonuclear generators. A study was done. A Mr. L.J. Ullian gave several of us (General Kutyna, Feynman, Armstrong, and me) a brief on March 11, 1986. He was with the Missile System Safety Division of the USAF section of DOD attached to the White House. According to his brief, and my notes, J.B. Baker of an organization called "Wiggins" of Redondo Beach, California, completed a study for the Air Force in 1981. The study, *Space Shuttle Range Safety Hazards Analysis*, Technical Report 81-1329, pp 3-31 to 3-32, compared the historical safety record of similar (to the Shuttle) solid rocket boosters such as those used on Minuteman, Poseidon, and AX missile systems during the previous two decades. Over 1,900 had been previously launched. The historical loss rate was one in fifty-seven.

However, the report also stated that, *Since the historical data includes motors developed 10 to 20 years ago, it was decided that a reduction in the failure probability was warranted for the Space Shuttle ... and to assume a failure rate probability of one per thousand.* The minimum main propulsion failure rate acceptable to NASA at the time for manned flight was 1 in 10,000. NASA's answer to this report was to provide an exhaustive statistical analysis (called the Sierra Report, ATWL TR 83-61 of December, 1983) that reviewed the Wiggins report. It was controversial, and the whole question went to the U.S. Congress and the Aaron Cohen Committee, which established an SRB failure rate of 1 in 100,000 launches. This is great. Just let the U.S. Congress establish failure rates. We who pay their salaries do not even know the educations of their staff members and most of the time themselves. This equates on average to, with daily Shuttle launches, one failure every 137 years. The historical failure rate of 1,900 similar solid rocket booster launches at the time of the Challenger launch was 1 in fifty-seven. NASA didn't accept this original probability of failure, and instead of helping to correct or reduce the risks, they helped change the criteria behind the projections. Basically NASA management provided themselves with blinders in order to ignore the real world of the technical environment they were managing. The actual failure rate of the Challenger-type solid rocket boosters at the time of the accident was 1 in fifty (two boosters per shuttle launch, twenty-five launches).

Radio Tape and On-Board Tapes

There were three possible types of audiotapes that could have shed light on the accident:

Personal audiotapes: NASA has always denied the recovery of personal tape recorders that could have been carried by crewmembers.

Radio tapes: NASA provided its transcript of the radio tapes (air to ground transmissions). This transcript was provided to the Presidential Commission during its tenure.

On-board Shuttle computer digital tapes: I also requested copies of all on-board tapes of Challenger. This request was met with the response that none existed; yet later the Coast Guard salvage team recovered one on-board computer tape. But NASA allowed no outside evaluation of the information on the tape. After this on-board tape (that had Challenger crew voices on it) was recovered from the ocean, it was taken to a lab to be refurbished after its immersion in salt water. NTSB sent their CVR expert, Paul Turner, to assist. I had been his backup for Cockpit Voice Recorder (CVR) analyses at NTSB for several years. Immediately after the tape was restored and several weeks before the end of the Commission tenure, two NASA employees arrived, showed identification, and took the tape. To this day no one outside NASA has ever heard this on-board audio tape recording. NASA released their version of a transcript of this tape after the Commission had disbanded.

It is important to note that all transcripts of audiotapes came from NASA; no outside person has been able to verify the completeness and accuracy of the transcripts. More importantly, Commission staff was unable to glean highly valuable information that may have been obtained from background sounds. Personal feelings and ghoulish accusations are not part of a professional investigation. I have heard dozens of recorded last words from pilots and crews about to die. But more often than not there is more information normally recorded on these tapes than voices. On airliners this often includes 400-megahertz background hums which can establish the health of the electrical power system, sounds of switch and lever noises, warning sounds like *whoop whoop, pull up*, engine sounds, and wind sounds. These sounds can be electronically filtered, digitized, isolated, and used like fingerprints. With the expensive sound equipment at NTSB alone, and the even larger FBI sound laboratory, there was ample government ability to analyze tapes if NASA had not sequestered them.

Hard Evidence of Crew Survivability

It was a normal launch. At about 73 seconds into it, there was an explosion that changed the flight direction of Challenger. Air molecules, beating against the vehicle at nearly twice the speed of sound, ripped it apart. But the crew capsule, built for the rigors of the complete vacuum of space, remained pressurized and intact. The crew knew something was wrong, and later physical information from the wreckage showed that they had accomplished at least part of their emergency checklist.

Part of this checklist included activating their emergency Personal Escape Air Packages (PEAP bottles). The commander and pilot could not activate theirs themselves since the bottles were behind their seats. Resnik probably activated Smith's. Four bottles were supposedly recovered and none have been examined by anyone outside NASA; but two were reported later by *Aviation Week & Space Technology* to have been partially depleted. With the types of valves (called "diluter demand valves") it is completely accurate to say that it is impossible that these bottles were partially depleted because of any explosion or accident impact damage. Someone had to turn the valve on. More importantly, only breathing could have partially depleted them. If the valve broke off during impact, the bottles would have been empty. If no damage had been suffered, the bottles would have been full. The only way the bottles could have been partially depleted, in this case by two to three minutes of use, was for someone to have been using them.

So, after the explosion, some of the crew was alive, thinking and breathing. Probably the commander and pilot were the only ones aware of the nature of their situation. As mentioned, had the crew been provided with small 5-pound sport parachutes for each, coupled with the parachute experience that the commander and pilot had gone through during previous required military training, they all would have had more than an even chance of surviving. But there they sat, with no options, watching the ocean arrive.

NASA Accident Contingency Plan

This concern is probably best summarized by a note early on to me from Astronaut Sally Ride. She wrote:

Jack - Is your panel looking into the adequacy/implementation of NASA's accident plan? My opinion (somewhat substantiated) is that NASA developed a pretty good plan in 1980 that

1. *Sat on the shelf & probably wasn't reviewed or updated often enough (if at all)*
2. *Wasn't* (her underline) *pulled out and used immediately after 51L (at least not by the managers who found themselves faced with the overall problem)* (her parenthesis)

I don't know whether this is something the Commission should look into, but I suspect it's been overlooked so far.

Sally

I looked into it after Ride's note. Actually there were two plans, neither of which was implemented. My take on it at the time was that neither plan would have saved anything or anyone. However, the fact that they were not used, when no one at the time knew what had happened, was indicative of NASA's attitude at the time. They weren't about to spend a nickel more on safety considerations. No wasted training time on "what ifs." The range safety officer, however, had a different and simpler set of contingency requirements. He was responsible to help ensure the safety of people on the ground. If major components of the Shuttle stack became unguided and could head for any populated area, he was both authorized and required to blow them up by means of coded radio signals to explosive shaped charges attached to those components. After the explosion he did blow up the solid rocket boosters, as they were then unguided and still thrusting.

Financial Motivations

NASA gained legal and binding settlement with the crew survivors before the Presidential investigation was completed. At this writing, although I still have a FOIA request filed with the Department of Justice, no one in the public sector knows if compensation was ever considered to include "conscious pain and suffering" and "wrongful death" of their loved ones. These two legal terms were paramount in NASA headquarters' mindset at the time since they may have led to huge legal monetary settlements. Without these two legal considerations, there would have been little reason to cover up or delay the release of hard evidence such as tape recordings, PEAP bottles, and other forensic evidence until a settlement had been made. NASA settled in secret with the survivors before the Commission came out with its report to the President. Prior to this settlement, Mike Smith's wife had sued NASA and Lawrence Mulloy for $15 million [see following page insert]. The general public does not know the result of this suit and its effect on the overall NASA settlement.

Personal Effects

The front end of the Shuttle actually consists of two layers or shells. On the outside, the part everyone sees is the outer shell with a separate windshield, small thruster hypergolic engines on the sides and nose for maneuvering in space, and part of the re-entry shield covered with heat-resistant tiles. Inside this shell is the pressurized crew compartment with living quarters, bathroom and shower, cockpit, storage, and the remote arm control station. Although they wore space suits, the crew did not need them unless pressurization was lost. And it was not impossible to raise a visor to talk into a recorder, or to hook a recorder into the crew's intercom system.

I requested any personal effects such as toothpaste tubes, hairspray bottles, or cameras that had been recovered, because such items could be hard evidence of explosive decompression and hence crew survivability. My request was met with stony silence from NASA.

JANE SMITH'S "CLAIM FOR DAMAGE, INJURY OR DEATH" FROM THE U.S. GOVERNMENT - ONE PAGE

CLAIM FOR DAMAGE, INJURY, OR DEATH

INSTRUCTIONS: Prepare in ink or typewriter. Please read carefully the instructions on the reverse side and supply information requested on both sides of this form. Use additional sheet(s) if necessary.

FORM APPROVED OMB NO. 43-R0097

1. SUBMIT TO:
National Aeronautics & Space Administration
and Lawrence B. Mulloy
George C. Marshall Space Flight Center
Marshall Space Flight Center, Ala. 35812

2. NAME AND ADDRESS OF CLAIMANT (*Number, street, city, State, and Zip Code*)
Jane J. Smith, Executrix of the Estate of Michael J. Smith, and the heirs of his estate

3. TYPE OF EMPLOYMENT
☐ MILITARY
☒ CIVILIAN

4. AGE
40

5. MARITAL STATUS
Married

6. NAME AND ADDRESS OF SPOUSE, IF ANY (*Number, street, city, State, and Zip Code*)
Jane J. Smith
1038 Shorewood Drive, Seabrook, Texas 77586

7. PLACE OF ACCIDENT (*Give city or town and State; if outside city limits, indicate mileage or distance to nearest city or town*)
Offshore from Cape Canaveral, Florida

8. DATE AND DAY OF ACCIDENT
Tues., 1/28/86

9. TIME (A.M. OR P.M.)
11:38 AM

10. AMOUNT OF CLAIM (*in dollars*)

A. PROPERTY DAMAGE	B. PERSONAL INJURY	C. WRONGFUL DEATH	D. TOTAL
	$100,000	$15,000,000	$15,100,000

11. DESCRIPTION OF ACCIDENT (*State below, in detail, all known facts and circumstances attending the damage, injury, or death, identifying persons and property involved and the cause thereof*)

See attached Exhibit A.

12. PROPERTY DAMAGE

NAME AND ADDRESS OF OWNER, IF OTHER THAN CLAIMANT (*Number, street, city, State, and Zip Code*)
Not applicable

BRIEFLY DESCRIBE KIND AND LOCATION OF PROPERTY AND NATURE AND EXTENT OF DAMAGE (*See instructions on reverse side for method of substantiating claim*)

13. PERSONAL INJURY

STATE NATURE AND EXTENT OF INJURY WHICH FORMS THE BASIS OF THIS CLAIM
Personal injury and Wrongful death - Plaintiff's testate was thrown about in the spacecraft and in the few seconds preceding his death, knew of his impending death.

14. WITNESSES

NAME	ADDRESS (*Number, street, city, State, and Zip Code*)
Several thousand people at the Kennedy Space Center, and untold thousands of others watching on television.	Names and addresses are too numerous to set out.

I CERTIFY THAT THE AMOUNT OF CLAIM COVERS ONLY DAMAGES AND INJURIES CAUSED BY THE ACCIDENT ABOVE AND AGREE TO ACCEPT SAID AMOUNT IN FULL SATISFACTION AND FINAL SETTLEMENT OF THIS CLAIM

15. SIGNATURE OF CLAIMANT (*This signature should be used in all future correspondence*)
Jane J. Smith, Executrix,
by W. F. Maready, Attorney

16. DATE OF CLAIM
July 2, 1986

CIVIL PENALTY FOR PRESENTING FRAUDULENT CLAIM

The claimant shall forfeit and pay to the United States the sum of $2,000, plus double the amount of damages sustained by the United States. (*See R.S. §3490, 5438; 31 U.S.C. 231.*)

CRIMINAL PENALTY FOR PRESENTING FRAUDULENT CLAIM OR MAKING FALSE STATEMENTS

Fine of not more than $10,000 or imprisonment for not more than 5 years or both. (*See 62 Stat. 698, 749; 18 U.S.C. 287, 1001.*)

STANDARD FORM 95 (Rev. 6-78)
PRESCRIBED BY DEPT. OF JUSTICE
28 CFR 14.2

Shortly after, or perhaps during the explosion, Mike Smith said over the radio, something like "oops!" or "uh oh," according to NASA's closed-door transcript interpretation. Phonetically there is a big difference here. Nevertheless, this final verbal recording, according to NASA, was just before the explosion, at about 73 seconds into the flight. NASA claimed no further conversation was recorded anywhere.

Later during the investigation I talked to Astronaut John Fabian about this. Fabian had been on two previous Shuttle flights, one of them in Challenger. Fabian said that it was not unusual in his experience for mission specialists or payload astronauts, some from other countries, to use personal hand-held recorders to tape their experiences during launch. Can one imagine a schoolteacher not doing this? Wouldn't she want to be able to later share with students her feelings during launch and first view of the Earth from above? Now 15 years later, a reputable aviation journalist, Dan Cook, told me that a private hand-held tape recorder was recovered shortly after the accident, its existence kept from the Presidential Commission, and that it was made by the schoolteacher during the launch and accident. It is still being subjected to a Freedom of Information request. Of course the contents of this tape would probably have had little or no import to the investigation. McAuliffe's words and impressions would be technically meaningless. However, audio background records of switch movement, decompression, and the words of technically aware occupants could have provided very important information. The important thing here was that this information was kept from the investigative body set up by the top representative, the President of the American people.

Crew Remains

Even the recovered remains of the crew were secretly taken at night from the salvage ships, in oil drums, to Patrick Air Force Base. There was no reason for any of the Commission or staff to be directly involved in this part of the investigation. However, there was a reason for a few on the staff to have access to autopsy reports of the Shuttle crew. Facts such as the oxygen levels in blood would have provided valuable information relative to decompression and levels of consciousness. Although I requested autopsy reports, and was told several times that they would be provided, none ever were.

NASA Obstructs Information Flow

During the first few weeks of the investigation, I talked to several dozen NASA engineers and requested much documentation, mostly technical. But NASA management stepped in, I believe, because I was dealing in large measure with NASA engineers who only wanted all the truth to come out, and were immediately providing me with all information requested. At its Washington headquarters, NASA soon set up a screening office and required all of their employees to send all information requested by the Presidential Commission through that office instead of sending it directly to the Commission. I took great exception to this and spent many hours discussing the issue with NASA and several commissioners. Here was a government agency being investigated that was controlling the information that went to the organization investigating them! To this day I don't understand why the level-headed Rogers allowed it. When Admiral Engen cautioned me not to let them waste my time, I hadn't realized how naive I was or how politically astute he was. As a taxpayer, I could not make any sense of this screening office.

The result of NASA's screening office was twofold—a delay of one to three weeks for requested information to be first approved by NASA, then perhaps given to the Commission, and a stonewalling of some requested information entirely. An example of this debilitating process was a request I made to get a copy of the crew's launch checklist. When I didn't get it I called the screening office and inquired. "Why do you want it?" asked the voice on the other end of the phone. I mentioned that it might pro vide useful information on the sequence of events. This screening office promised this information, but never provided it, even after many telephone calls by me.

Crippen

One large piece of new wreckage reportedly had the American flag painted on it. I decided to inspect it and get some photographs. So I flew down to Canaveral and went to the hangar where the spacecraft wreckage was kept and began taking photographs. Soon a guard scurried over and told me, “No pictures without a permit.”

NASA Astronaut and Navy Captain Bob Crippen was in charge of the Canaveral part of NASA’s participation for the Commission investigation that included wreckage and people recovery from the ocean floor. It is important to note that Crippen was not part of the Presidential Commission or its staff and had no authority over them. So I told the guard to call Crippen and verify that a photo permit was not required.

A few minutes later Crippen called. I taped my half of the dialogue. The tape may have gone to the NationalArchives along with my notes. I don't have them. But the episode was so outrageous that it burned into my mind. My recollection of the conversation is as follows:

Crippen: *“Jack, I don’t want you taking pictures.”*

Macidull: *“Yesterday you had a hundred newsmen in here taking thousands of pictures, and they are not even part of the investigation. What gives?”*

Crippen: *“Well, that was not my idea. We have our own camera people to take pictures.”*

Macidull: *“Bob, I’m down here to conduct an investigation. There is only one person between me and the President of the United States. If you have a problem, talk to Mr. Rogers. In the meantime I’ll be here taking pictures.”*

Crippen: *“We have a photographer who can take all the pictures you want.”*

Macidull: *“I know about your photographers. Tell you what. My next stop is the SRB and ET wreckage hangar. If you have a photographer there in 15 minutes who can take photographs in two-second intervals instead of the usual one minute, I would be willing to use him.”*

After the call, I remembered a day at Canaveral a month before when I was waiting to board a military plane to fly back to Washington. There was a sleek NASA supersonic T-38 jet landing. It taxied up to the small Cape Canaveral Operations building. The pilot shut down the engine and out of the cockpit stepped an astronaut, one not scheduled to go into space again. He walked across the flightline and went into a room in the Ops building—which included a pilot ready-room with private lockers. Though he was a military officer, his locker contained a civilian suit and an expensive pair of shoes. Once changed, he looked more like a corporate CEO than an astronaut who had courageously helped pilot one of the first space Shuttles when they had ejection seats and parachutes, and again later when the Shuttle had neither. He left the building and went to the parking lot, where he got into a civilian rental car and drove off. Where had he been? I checked with operations and found out that he had been to Houston for the weekend. There was nothing wrong with keeping flight skills up by flying on weekend junkets, but he was not scheduled to go into space again.

His clothes, private locker with no uniform, and the free NASA jet were all signatures of a "club member." His flight that day with the fuel, maintenance, high-time component replacements, and ground support equipment and personnel, in 1986 dollars cost the taxpayers about $10,000. Then, it would have cost about $200 to fly commercially from Houston to Canaveral.

At the SRB hangar, a NASA photographer was there and took many pictures at my request. But when I pulled out my camera, my escort, a female NASA astronaut assigned to me, said I was "violating national security!" I let her stew as I snapped away ... "violating national security." I was on orders from the President as part of the Commission. I was beginning to wonder who the hell these people thought they were and why they were trying to cover everything up.

That evening Crippen called me at my motel room and asked me to surrender the film of the pictures I had taken. Surrender? Was I the enemy? This seemed both curious and puerile. I said I had pictures of my girlfriend on the roll and would rather have them developed myself. Crippen insisted that NASA would develop the film and only keep copies of the Challenger shots. As a show of faith and trust in the common cause, I said I would loan my film to NASA for development. Later that evening a NASA employee showed up at my door to collect the film.

The next day, I was presented with the developed film and negatives. I opened the envelope and there was my girlfriend. But there were no other prints. And NASA had cut off and kept all the other negatives. But I had given them only one of my two rolls taken of the wreckage. Several of the pictures in the second roll were used in the final report, including the one with the American flag on the side of the Challenger.

More NASA Gamesmanship

After the Presidential Commission was disbanded, NASA released a study that said the evidence was "inconclusive" as to whether the crew survived. The crew survived the explosion; we know that now. NASA knew at the time because there was hard, conclusive evidence to prove it. So why all this confusion and continued intrigue about such evidence? Why wouldn't NASA rush to embrace the truth?

It cannot be stressed too strongly that NASA was going through a metamorphosis, from a public agency that wanted everyone to love manned space flight to a covert agency that demanded that nothing be leaked to the media. That's a 180-degree flip-flop in public policy. Why did NASA wait to reveal what they said was inconclusive evidence? Scores of NASA engineers and some managers I talked with during the investigation were completely open and helpful in providing the Commission with any information requested. They often volunteered more information than the Commission requested or didn't know enough to ask for. When NASA established its screening office for all information designated to go to the Commission, they held some things back. After the Commission was disbanded, some of these engineers and managers, having read the report to the President, realized that their information hadn't gotten through. They made noises to the effect that concealed facts would be made available to the public in other ways. Much of this effort was directed to the media and the publication, *Aviation Week and Space Technology*.

During this investigation, NASA was consistent concerning the media. If headquarters was threatened with release of information, they immediately released it themselves to avoid being accused of a cover-up. Certainly there was a reason for this tendency to cover up certain evidence—a legal reason—a money reason that also had to do with time.

CHAPTER VII
The Challenger Legacy
By Macidull and Blattner

This book is the result of facts obtained after more than a decade of effort to obtain useful information and provide the story of the Challenger accident from the perspective of someone who lived the accident investigation. This book was a painstaking job; but the groundrules were simple. I attempted to allow only validated facts and testimony with few suppositions or personal opinions except in this final chapter (VII) and the appendices. With this attempt I failed. Personal views and opinions have appeared throughout this history, both by my co-author and me. However, in this work it is doubtful that the authors' opinions will become a matter of libel or slander under the laws of this country. The relative causes, processes, dialogues, laws, and regulations (which are a matter of record) are contained in the first six chapters and appendices and reveal much of what actually happened during this tragic event. With this final chapter I attempt to separate things with a freedom of opinion by both Mr. Blattner and myself. And to paraphrase Plutarch, *"As painters in their portraits labor the likeness in the face, particularly about the eyes, in which a particular turn of mind most appears, and run over the rest with a more careless hand, so we must be permitted to strike off the features of the soul, in order to give a real likeness."*

Some Final Comments From Les Blattner

Writing about Challenger in the abstract some 16 years later brings to mind Carl Sandburg's statement: " ... the past is a bucket of ashes." Sifting through the ashes of the Challenger accident — ashes that have been stirred and repeatedly picked through by countless reporters and authors in books, articles, and documentaries—has been a daunting task.

At this writing, the Challenger accident is over 16 yeas old; and in more than a decade, some of the facts have fallen victim to revisionist history. Take for example the major belief by the public that the Challenger accident was about O-rings. "Why didn't they fix that problem before the launch?" was a common question encountered during the writing of this book.

But after more than a year reading and rereading testimony and comments made during the Presidential Commission to investigate the Challenger accident, there is only one question I want answered. Why did the flight of Challenger take place at all on that cold Florida day? In studying the documents amassed for the investigation, one is struck that the seven astronauts aboard did not need to lose their lives on January 28, 1986. My feeling is that, had this been a Project Mercury launch, no life would have been lost. In the 1960s, engineers would never have allowed Challenger to fly beyond its operational limits.

And that brings us to the real problem. Why was Challenger launched on that day when Thiokol engineers said flatly that the temperature was beyond the operational limits previously tested for solid rocket boosters used in live crew launches?

Thiokol knew there was an O-ring problem. They knew it early in 1982 when tests turned up the O-ring anomaly in burn-through during static tests in Utah. By August 1985, engineers from Thiokol wanted a high-level meeting with NASA. In that meeting they laid out the problem and said it should be fixed. They told people such as Jessie Moore, NASA's Associate Administrator for Space Flight, about the O-ring problem in specific terms that the Shuttle should be grounded until the problem was fixed. "No one could have left that meeting without knowing there was a problem," said Thiokol's Alan McDonald during our interview. For their trouble, NASA's only public response was that Thiokol was trying to get the agency to pay for a new solid rocket design.

But as stated, the O-rings were not the reason seven astronauts died in Challenger. Let there be no illusions or gray area here—the Challenger disaster was *a man-made error*. Or, put another way, there was a lack of accountability. While there were procedures and rules by which launch decisions were made, several of NASA's launch mangers did not follow these rules, according to Commission testimony.

While the reporting of information to higher levels was addressed in a Commission recommendation, the accountability issue was certainly not solved. Let me digress slightly because NASA has an institutional problem in being honest with the American public.

In 1969, as indicated in Appendix III, NASA tried hard to convince the Nixon Administration that a manned Mars landing was not only feasible, but also necessary. Remember that NASA was created in the cold war to place a man on the moon. After this mission was over, the space agency's enabling language did not give NASA a new mission, other than in the broadest terms. So after Apollo, NASA was desperately trying to reinvent itself to Congress.

When the bottom fell out of the NASA budget after Apollo moon missions, a Mars manned mission was floated as the next step into space. But when the Nixon Administration refused the idea, the Shuttle program was grasped as the next evolutionary step into space.

NASA started mapping Mars long before Challenger flew. That began with Mariner 3 launched on November 5, 1964 from Cape Canaveral. After a short delay because of some Agena booster rocket circuitry problems, the launch went normally; but an hour later, telemetry indicated that, while the scientific instruments were functioning, there was no power from the solar panels. The first Mars probe blindly headed out into space, destined to enter a solar orbit.

Of course, that was only the first such launch; and there have been numerous successes along with a few failures in programs such as Mariner, Viking, and the various other landers going to the Red Planet. Still, that Mars launches continue today, unabated, four full decades after the Nixon administration said no to men going to Mars, with a price tag of billions of dollars, is amazing.

Yet, in 2001, the agency's lack of accountability for the Challenger crash astonishes not only me but also everyone who ever wanted finality to this tragedy.

The question begged down through the years remains: why wasn't someone held accountable for this senseless loss of life and hardware? Neither the President, his commission investigating the accident, Congress, or NASA ever pointed a finger at any of those responsible.

Without question, the American public had a right to know the answers about the Challenger accident. The U.S. taxpayer paid for the program, and didn't they deserve to know? Of course we did— but not 16 years later, not one year later. We wanted to know within days of when the accident happened, when it would have mattered. Didn't we deserve the truth? Didn't the guilty deserve to be punished?

The President charged the Commission to review the circumstances surrounding the accident and establish a cause or probable causes. So of course, engineering reasons for the accident were the target. Therefore, on June 6, 1986, when Chairman Rodgers sent his report to President Reagan, the Commission assumed it had accomplished that goal.

That the report fell short of specifically naming those responsible for the decision to launch Challenger is no mystery. The enabling language from the President never specifically said *determine who was guilty*. Chairman Rodgers had a blank check to determine the mechanical failure, and deal with some periphery issues such as criticality. But as Macidull puts it, "It's impossible to convict O-rings of a crime."

Reading through the transcripts of testimony of NASA managers who testified under oath before the Commission, there comes a realization that the NASA managers worked hard to obfuscate the truth. That's not to let the Commission off the hook. The commissioners had these men squarely in their cross hairs but couldn't pull the trigger. Would it have mattered anyway? There was no political resolve and the commissioners had no ammunition for their guns.

Other ironies surrounding the investigation began with the FBI agents named to work on the Commission. Their presence sent a message that there was some criminal wrongdoing. Certainly there was, since the decision to launch Challenger was subjective and made by a few people covering up the facts.

Perhaps the greatest irony of all was that the Commission did get to the bottom of the problem. In its final report there is hard evidence in quotes about the wrong-doing in Florida. In Volume I, page 104 to 110, there is a running dialogue about the events that tell the climax of the terrible tale. The Commission heard this testimony. They pieced together the evidence. But that's all they did.

After going through the Commission testimony and talking with others who were in Florida during the launch, one can only conclude that NASA managers such as George Hardy, the Deputy Director of Science & Engineering at Marshall, and Stanley Reinartz, Marshall's Manager of Shuttle Projects, were responsible for institutional manslaughter. They were not the only guilty parties, but they certainly head the short list.

On the pages of the final report, as in real life, Hardy and Reinartz argued and fought to launch the Challenger. They did so, in spite of engineering reasons at Thiokol that said not to launch in the freezing temperatures that day in Florida.

This would have been a no-brainer for Project Mercury engineers. But their own beliefs and desires drove Hardy and Reinartz. When I interviewed James Beggs, the former NASA administrator, he told me there was no written or spoken policy demanding managers to launch any Shuttle mission until everything was proved safe—period. Certainly after Apollo, NASA testified before Congress that the agency would launch space Shuttles like "clockwork." But Beggs was adamant that the agency wanted safety first, and no one forced a Challenger launch because of a need for a quick turnaround.

Still, if there was no requirement to turn the Shuttle around in two weeks, why did Hardy and Reinartz argue so vehemently to launch? Was it pride, ego, or anger that made them do it? As one reads their testimony in the final report, the conclusion cannot help but be reached that these two managers served as NASA enforcers to make sure that Challenger was launched ASAP. But why did they risk human life to do that? Were these men breathing the intoxicated air of government hierarchy not accountable to anyone?

There was Alan McDonald, Morton Thiokol's Solid Rocket Motor Director, who took on both managers and adamantly presented his engineering and personal concerns, only to be slapped down like some inconsequential fob. In my eyes, McDonald was a hero in the Challenger drama. He thought he would be fired because of his opposition was too vehement with Thiokol's largest customer. Still he did try to stop the launch, even though Thiokol senior management subsequently sold him out.

A large portion of Commission testimony focused on Lawrence Mulloy, Marshall's solid rocket booster project manager. Indeed Mulloy did help to confuse and mislead the Commission. His bewildering comments concerning the O-ring seals and the *Criticality Issue* before the Commission defies believability. Perhaps in this respect he should have been held accountable. But he reported to Reinartz, and ultimately it was Reinartz who wanted, yes, insisted on the launch.

Then there was Jesse Moore, NASA Associate Administrator for Space Flight at NASA Headquarters. In front of the Commission, Moore denied any knowledge about the O-ring discussion or any problem the morning of the launch.

Covering NASA in 1960 as a cub reporter, I remember thinking that top managers were the best and the brightest. Yet in front of the Commission, Moore said publicly that he did not understand the Criticality Issue, and the public record shows that he never attended a single Shuttle preflight briefing.

Throughout his testimony, Moore seemed almost ambivalent about the Challenger launch. Did he truly care that Thiokol had been telling NASA, in meetings he attended, that there was an O-ring problem since 1985? Certainly he was involved and aware of these facts. Where was his leadership on this issue? He voted to launch Challenger, so shouldn't he have been held accountable for his poor execution of duty?

The Challenger accident cost our nation dearly in money and human resources. The single largest accident investigation in the history of the world! A threat to national security! A setback to the scientific community of many decades! And, for all of this, no person or parties were responsible?

By the time the Commission met, almost everyone knew that the Thiokol booster solid rocket failed. Most had watched it on television, sometimes in slow motion. Did we need a presidential commission to tell us that it happened?

Half the Shuttle missions flown in front of Challenger had a similar or related problem to the one that caused the accident.

Wasn't that enough engineering proof to hold up the program until a fix was implemented?

Obviously the accident wasn't simply NASA's fault. Morton Thiokol had its own share of blame to shoulder. Alan McDonald told me that once, during a recess in the Commission hearing that he and Roger Boisjoly, an engineer on his staff and member of an internal panel to investigate the seal problem, were approached by a congressman who asked if Boisjoly was so convinced that rocket would fail, why didn't he call the White House? Boisjoly replied, according to McDonald, that he never considered such an action. But the question is a valid one. If there was such concern about the solid rocket motor O-rings performing in the cold weather, and NASA wouldn't stop the launch, and your own senior management began to waffle, why didn't someone from Thiokol call a congressman, senator, or the President?

With the lives of seven people, and nearly two billion dollars of hardware riding on the outcome, Commission testimony shows that Thiokol, after debating the issue for two days, simply folded its tent and acquiesced to appease its largest customer. Even now, it seems incredulous that the nation's premier solid rocket manufacturer, which based it success on specific scientific principal, gambled that the rocket motor would work.

As per Commission testimony, the architect of such a tactic was Jerald Mason, the Wasatch vice president who implored other managers to "take off your engineering hats and put on your management hats." This statement confirms that high-level managers leveraged the deal that allowed Thiokol to hedge and then ignore its engineering knowledge.

Of course, Mason couldn't have pulled off the feat without the support of Robert Lund, Thiokol's Vice President of Engineering. According to Boisjoly's testimony to the Commission, Lund, like a large percentage of rank and file engineers, favored calling off the launch. This was another point the Commission never clarified. To reverse his engineering conclusion must have been a humiliating blow to Lund. But if it was, he apparently didn't fight it. Lund fell into line like a dutiful corporate solider, and was never heard from again during the Challenger launch deliberations.

Thiokol, for all its part in the death of the astronauts, was rewarded with a $480 million redesign of the solid rocket motor, and subsequent NASA and military contracts.

Yes, the Challenger accident was preventable. Yet even after all these years, it remains curious that no one blew the whistle, especially with that question about NASA management. The irony was that NASA supported the Commission. There were NASA astronauts on the Commission staff. There was a new NASA administrator when the Commission was created. The Commission was paid for out of the NASA budget. Even had there been an attempt to investigate NASA for its part in this accident, it would have been as if NASA were investigating itself.

After 120-days of investigation, and a bouillabaisse of comments and recommendations, the Commission went out of business. And finding out who was responsible for the Challenger accident was lost to antiquity. The political process had done its work. The President was off the hook, NASA was cleared, and Thiokol went on to become more profitable and later sold to Cordant Technologies.

It has been pointed out that in the final minutes of countdown, no one voted to stop the launch, not NASA, not Thiokol. Of course this is correct. Hardy Reinartz and Moore, as NASA top government management, and Thiokol's own senior managers Mason and Lund, worked to stifle any opposition. Had Challenger succeeded, it would have been just another day at the office. In failure, however, the shroud was lifted and the spin and illusion removed.

Should these men have been held accountable? If not by the Presidential Commission, then by Congress—if not by Congress, then by the President of the United States—if not by our elected leaders, then by the public. Didn't anyone truly care that seven people lost their lives and they called it an accident?

Some Final Comments from Macidull:

During and after the accident, and I use the word *accident* reservedly—as in possibly replacing it with the term *acts of criminal negligence*—millions of school children watched it on television. Their parents watched it also. There was no authority vested with the National Transportation Safety Board (NTSB) for accidents relative to space flight. So President Reagan had to set up a commission. This commission of thirteen members and a staff director had no education, background, or experience in aircraft had no education, background, or experience in aircraft accident investigation. Except for Armstrong, the Commission members were selected without engineering thought, experience, or in the light of any public scrutiny. They were simply the commissioners appointed by the President. Fortunately for the investigation, the head of the Commission, William Rogers, did have experience in politics and a clear head to recognize the gap between him and the other commissioners. He also recognized after two weeks that the current situation would not work. There was not even a plan.

There were two astronauts on the Commission staff who were not only "in-house" with NASA but also had no experience in accident investigation. During the third week Chairman Rogers didn't even know where the commissioners were. So he called the Secretary of Transportation, Mrs. Dole, requesting technical help. She knew nothing technically and so referred him to the FAA Administrator, Admiral Engen, who called me. I did my best during the Commission tenure and received several accolades. It was hard work. At times I felt I was dealing with spoiled children, or trying to herd cats. But the specified job got done, a job only couched in Executive Order number 12546 from the President. It stated that the members of the Commission *shall be drawn from among distinguished leaders of the government, and the scientific, technical, and management communities.* Nothing was mentioned about aircraft investigative communities, which included professors of accident investigation from many colleges around the country, the military, and the NTSB. Who determined who was distinguished? Who determined what constituted a leader in those communities? And what about the other provision in the President's executive order?

"Notwithstanding the provisions of any other Executive Order, the functions of the President under the Federal Advisory Committee Act which are applicable to the Commission, except that of reporting annually to the Congress, shall be performed by the Administrator of the National Aeronautics and Space Administration, in accordance with guidelines and procedures established by the Administrator of General Services."

Were these guidelines regulations? Were these undefined procedures also considered regulations which had the same weight of law as in other government agencies?

So we had an investigation established by an executive order where there was no requirement for professional aircraft accident investigators, where the "*functions of the President...shall be performed by the Administrator* (of NASA)," the very organization being investigated where the determination of crew survival was not required, and where the task was required to be completed within four months. As stated previously, major airline accidents take a minimum of a year at NTSB with dedicated laboratories and trained personnel in the fields of wreckage analysis, radar and audiotape analysis, weather analysis, metallurgy, operations, aircraft performance, flight simulation, data reduction analysis, human factors, and legalities.

The Commission was therefore hamstrung from the beginning. What the public got was a very good report on the hardware "cause" of the accident—the O-rings, not the real cause. The O-rings had been failing for years. In 1982, they were supposed to be fixed before further flight took place. NASA management ignored its technical and legal regulations. The real cause of the accident was the group of people involved in the decision to launch under untried environmental conditions with a known defect. This was the same defect susceptible in the past to similar environmental conditions; but in this case it was more severe and against their own required launch criteria, which took many decades to develop. Though they were required to adhere to those launch criteria in their high-paid positions of public trust, they did not. As stated before but repeated here for emphasis, the head of Shuttle launches never even bothered to show up for any of all the previous twenty-five pre-launch meetings where potential problems were discussed by many concerned NASA engineers. Although during public hearings the Commission addressed some of this, it was all eventually swept under the rug. Four months does not allow time for a trial. O-rings cannot testify. The political and media atmospheres were to get it over with and to move on. This was the convenient way to do it for all responsible—without accountability.

During and after the investigation, nothing came of NASA's flagrant obstruction of the Commission investigation, cover-up of critical information, and the withholding of evidence. The Commission staff requested much information that was not provided—much of which was later found to exist. Ironically, NASA was not legally obligated to fully cooperate. The truth had already been sequestered by the President's executive order—first, get people who knew virtually nothing technically about conducting such an investigation, then require that the agency (NASA) being investigated provide the funding (dollars previously approved by Congress for other purposes) and then put their own people in to help investigate themselves.

However, the truth has come out anyway. It just wasn't brought to the NASA owners' attention, namely the U.S. taxpayers, largely through the media. Perhaps it was too difficult and untimely to put together for public consumption. Perhaps most just weren't interested anymore, according to the media. Perhaps in the minds of many there were no lessons to be learned from reopening a can of worms. This author's opinion is that management was responsible for the death of these seven astronauts and the destruction of all the equipment involved, blindly, without cause, sending everything aloft without regard to known system failures relative to known environmental conditions. These managers acted, in my opinion, like spoiled, uncontrolled, undirected, and unadvised children who had little educational or personal experience compared to those they sent. And so they killed people.

This book also contends that the media does not control all of what the public may want or need to hear or know, but that it is in the public's interest to have access to all of what their tax dollars bought, like the educations and experiences of managers, and then decide for themselves. Perhaps then, our type of government could avoid having this type of disaster repeated.

APPENDIX I & I-A (also in the U.S. National Archives)
I - 2/23/86 PLAN FOR CHALLENGER INVESTIGATION

I-A - 3/10/86 PRESIDENTIAL COMMISSION OPERATING PROCEDURES
By Macidull on February 23 and March 10, 1986

++

Appendix I

PRESIDENTIAL COMMISSION ON SPACE SHUTTLE CHALLENGER ACCIDENT

PLAN OF INVESTIGATION

John C. Macidull
Investigator
Presidential Commission
February 23, 1986

SUGGESTED METHODOLOGY

Attached are four general categories of endeavor, parts of which should be added to and/or deleted as the investigation progresses.

PRELIMINARY LIST OF REQUIRED INFORMATION
PRELIMINARY SUMMARY OF STAFF DETERMINATIONS
PRESIDENTIAL COMMISSION FINAL REPORT INCLUSIONS
CALENDAR MILESTONES

These categories can be used to schedule and assign activities relative to the investigation. They are not strictly sequential in time, but can be used to view and guide overall progress.

A master log should be kept on all information received and requested. This information, upon receipt, should be labeled as to which staff determination it addresses, and kept in a specific location, or library, when not in use. If data on microfiche is used, a hard-copy machine should be obtained.

One or two word processing machines with letter quality printers should be acquired immediately.

The list of staff determinations should be reviewed by the Commission as soon as possible.

Assignments of determinations should be made to staff members in accordance with their background and experience. Staff members are encouraged to write out their findings as they go along, and have them entered into a word processing system. The secretariat should keep an ordered file in print on the latest versions of these inputs.

A final report outline should be written during the first thirty days.

Small groups should be formed in about sixty days, which include both staff and Commission members, for the purpose of addressing the final report inclusions. Each group should have a writer.

A month before the final report is released, any visual aids, diagrams, etc. needed for the report and/or formal presentation to the public, should have been contracted out.

A draft of the final report should be circulated and reviewed, for correction and editing, by all Commission and staff members at least two weeks prior to release.

PRELIMINARY LIST OF REOUIRED INFORMATION

1. All pertinent manuals and schematics for:
    a. SRB's
    b. Centerline external fuel tank
    c. Vehicle (Challenger) fuel, control, and guidance systems
2. A listing of all TM data available during this and previous launches.
3. Videos of all previous launches from all angles.
4. NASA accident launch communication flow diagram with names and phone #s.
5. NASA accident launch/abort decision block diagram with names and phone #s.

6. A list of all Thiokol–NASA interface people with phone #s.
7. Bios on all middle- and top-level accident launch decision makers.
8. A list of all command destruct subcontractors, system schematics and interface people and phone #'s
9. A list of the NASA Shuttle cryogenic experts and phone #s.
10. A list of he NASA structures experts for the vehicle, SRB's, and external fuel tank, with phone #s.
11. A sequential list of vehicle, SRB, and centerline tank modifications from launch 1 to 25.
12. A time-after-launch listing of all abort scenarios and procedures (non-survivable abort, ocean ditching, alternate airfields, etc).
13. All transcripts of testimony relative to this investigation.

PRELIMINARY SUMMARY OF STAFF DETERMINATIONS

Technical

1. SRB abnormal fire location and cause.
2. Effect of abnormal fire on centerline tank.
3. Centerline tank destruct system activation.
4. Centerline tank liquid hydrogen detonation.
5. Effect of adverse SRB thrust correction causing failure of attachment structure.
6. Jet stream turbulence encounter and relation to high-Q dynamic pressure.
7. Effect of full throttle transition. Possibility of centerline tank fuel pump failure.
8. Centerline tank destruct system radio frequency/code activation by extraneous source and subjectivity to sabotage.
9. SRB solid propellant shrink/cracking as a function of low temperatures at the time of the launch.
10. Effects of SRB abnormal flame propagation as a function of propellant crack widths and depths.
11. Survivability of crew compartment within the vehicle explosion but before water impact.

Human

1. The communication process inside NASA in relation to the launch.

2. The communication process between NASA and the outside in relation to the launch.
3. The threshold positions in NASA where launch-pertinent communication can be blocked from ascending further.
4. The identification of any critical information that was not forwarded past threshold positions.
5. The identification of any human pressures that influenced judgement during the launch decision process.

PRESIDENTIAL COMMISSION FINAL REPORT INCLUSIONS

1. The sequence of physical and environmental events
2. The sequence of human events
3. The time correlation of physical/environmental events to human events
4. The effect of physical/environmental events on human events
5. The identification and effect of critical information availability to key persons in the launch sequence
6. The findings and conclusions of the investigation
7. The causes and factors contributing to the accident.
8. Recommendations

CALENDAR MILESTONES

February 25: Technical investigative and writer roles established. Staff determinations reviewed and finalized. Calendar expectations relayed to media.
March 22: Release of questions (Staff Determinations) of which we are seeking answers, to the media. Estimate of final report date. Final report outlined.
April 19: Release conclusions on some preliminary findings.
May 17: All required information received and categorized. Begin final draft of report text.
June 14: Final report released.
(End of Appendix I)

APPENDIX I-A

PRESIDENTIAL COMMISSION ACCIDENT ANALYSIS PANEL

INVESTIGATIVE PLAN AND OPERATING PROCEDURES

Jack Macidull
March 10, 1986

VOCABULARY AND ITEMS TO NOTE

• FAULT TREE: A block diagram which flows from left to right and begins with the event of interest (in this case the explosion), and branches out into the possibilities of cause and contributing factors. The Accident Analysis Panel addresses only physical items of possibility. It is necessary to eliminate elements in the Fault Tree as far to the left as possible.

• ANOMALY: Something which is irregular or abnormal. Its application here is as an observed or measured deviation from a standard or normal condition. The Accident Analysis Panel has requested from NASA a list of all known physical anomalies without rank as to relative importance.

• SCENARIO: A hypothesis of a chain of events which correlate the known anomalies with the Fault Tree and which lead to the explosion. Scenarios thread through the Fault Tree from right to left and end at the event of interest.

• TIME LINE: An accepted chronology of events. The Accident Analysis Panel is concerned with a Time Line from SRB ignition to the right SRB Range Safety Package detonation.

• The resources at the disposal of all Commission Panels are nearly limitless for the next few months. To utilize them however, will require a high percentage of delegation, often to more than one organization for the same effort, and a system of tracking actions and evaluating results. Until such time as an outline for the final report is available, the panel titles and areas of concern should be used as a guide for writing preliminary drafts.

INVESTIGATIVE PLAN (Accident Analysis Panel)

The general plan of the accident analysis part of the investigation is to:

1. Determine, acquire, and review all pertinent and source material.
2. Ensure that all appropriate tests are determined and completed in a timely manner ('timely' to be determined by Armstrong and me, as was later the case).
3. Address every applicable event and hardware item that could have influenced and/or contributed to the explosion in the Accident Analysis Panel's purview.
4. Interview or arrange transcripts of interviews of all parties who can provide pertinent knowledge and experience relative to events leading up to the explosion.
5. Establish and update weekly a central master list of required Panel actions, i.e., evaluations, travel, interviews, written summaries, etc.
6. Utilize the Panel Operating Procedures (included on the following pages) to establish a most probable physical cause of the accident, and any factors which may have contributed.
7. Utilize test results, expert testimony, and Panel member experience to provide various recommendations that will enhance future Shuttle safety.

Attachment I-A OPERATING PROCEDURES (Accident Analysis Panel)

The procedures for Panel operation will be largely driven by the logic of the Fault Tree/Anomaly/Scenario troika. The collation and interrelationship between and among the elements in these investigative devices will establish what most probably happened physically, which resulted in the accident.

The point of departure for the Accident Analysis Panel is the acquisition of a complete Fault Tree and list of anomalies. The Fault Tree that was provided by MSFC is not in the logic format required, and another should be requested. A list of anomalies was requested last week.

Given the above two documents, Panel members should address each element in the Tree and cross-reference them to the list of anomalies. Some elements will branch into other elements which will require special tests, evaluations, travel, and interviews, etc. The Panel leader should assign these elements which require special attention to specific Panel members, who would be responsible for providing a list of source material and a short written summary which colors the element probable, possible, or unlikely.

Panel members will request information and assign work, studies, tests, and special evaluations, etc. through the Commission's 'action tracking' center, i.e., Emily Trapnell. She will formalize all requests from the Commission, and provide whatever is requested when it arrives.

When the Panel member has completed 'coloring' the assigned element, a sign-off summary sheet with his/her conclusions will be attached to the front of the supporting document(s) and forwarded to NASA Headquarters (Truly/Honeycut) for their comments and sign-off (again, through the Commission's Action Tracking office). When it is returned, if there is agreement, and no additional information is to be considered, then that element will be closed out.

After all the elements on the Fault Tree have been closed out, and all the anomalies properly addressed, scenarios should be written which include reference to the pertinent anomalies and the Time Line of events.

I will maintain an updated master version of the Fault Tree and list of anomalies.

Panel members should be involved in any element they desire, but only responsible procedurally for the ones assigned. O-ring and field-joint questions, for example, should be addressed by all Commissioners, but not to the degree that excludes their evaluation of 'minor considerations' that are critical to all thorough investigations.

Although there is a natural tendency to focus on the most obvious physical cause of the accident, and all are encouraged to provide input in that direction, a complete investigation requires uncovering and addressing all aspects, and to include or eliminate all the possibilities that could either have contributed to the accident, or could influence future practices, lines of communication, and hardware design.

End of Appendix I-A

APPENDIX II

CHALLENGER COUNTDOWN and LAUNCH SEQUENCE
By Macidull

The following is a record of the time sequence of events according to (1) the Shuttle-to-ground radio transcript, (2) on-board computerized crew recording transcript, (3) instrumentation readings received from the Shuttle by telemetering equipment on the ground, and (4) correlated ground photography. Entries citing "From Presidential Report" are sourced from the referenced Shuttle-to-ground radio transcript.

This sequence of events has never been collated prior to this book. In addition, I have attempted in near-lay terms to explain the technical nature of what transpired during the launch and accident sequence based on all the information so far available. The letter "T" signifies clock times before (minus) and after (plus) the number of minutes and seconds relative to actual liftoff, which is T minus 0.00 seconds. These times are relative to the liftoff time of 10:38 EST, January 28, 1986. NASA descriptions of crew comments are from the on-board recording equipment that were only made available after the Presidential Commission had written and published its final report to the President.

T minus 2 minutes, 5 seconds: (From NASA transcript of on-board voice recording) Mission Specialist Judith Resnik said, *Would you give it back to me?*

T minus 2 minutes, 3 seconds: (From NASA transcript of on-board voice recording) Mission Specialist Judith Resnik said, *Security blanket.*

T minus 2 minutes, 2 seconds: (From NASA transcript of on-board voice recording) Mission Specialist Judith Resnik said, *Hmm.*

T minus 1 minute, 58 seconds: (From NASA transcript of on-board voice recording) Commander Dick Scobee said, *Two minutes downstairs; you gotta watch running down there?* NASA description of comment: "Two minutes till launch."

T minus 1 minute, 47 seconds: (From NASA transcript of on-board voice recording) Pilot Mike Smith said, *OK there goes LOX arm.* NASA description of comment: "Liquid oxygen supply arm to ET." Macidull's comment: The liquid oxygen (LOX) tank is the topmost part of the Shuttle stack. Since liquid oxygen evaporates quickly through a pressure relief vent, it is topped off from an oxygen nozzle connected to a mechanical arm attached to the launch pad gantry. This helps insure that a full load of liquid oxygen is available for launch.

T minus 1 minute, 46 seconds: (From NASA transcript of onboard voice recording) Commander Dick Scobee said, *Goes the beanie cap.* NASA description of comment: "Liquid oxygen vent cap." Macidull's comment: The oxygen tank vent cap/nozzle on the nose of the brown external tank (ET) is disconnected and swung away.

T minus 1 minute, 44 seconds: (From NASA transcript of onboard voice recording) Mission Specialist Ellison Onizuka said, *Doesn't it go the other way?*

T minus 1 minute, 42 seconds: (From NASA transcript of onboard voice recording) ***Laughter in cabin.***

T minus 1 minute, 39 seconds: (From NASA transcript of onboard voice recording) Mission Specialist Ellison Onizuka said, *Now I see it; I see it.*

T minus 1 minute, 39 seconds: (From NASA transcript of onboard voice recording) Pilot Mike Smith said, *God I hope not Ellison.*

T minus 1 minute, 38 seconds: (From NASA transcript of onboard voice recording) Mission Specialist Ellison Onizuka said, *I couldn't see it moving; it was behind the center screen.* NASA description of comment: "Obstructed view of liquid oxygen supply arm." Macidull's comment: Reference to the launch pad gantry arm that capped the external tank oxygen vent.

T minus 1 minute, 33 seconds: (From NASA transcript of onboard voice recording) Mission Specialist Judith Resnik said, *Got your harnesses locked?* NASA description of comment, "Seat restraints."

T minus 1 minute, 29 seconds: (From NASA transcript of onboard voice recording) Pilot Mike Smith said, *What for?*

T minus 1 minute, 28 seconds: (From NASA transcript of onboard voice recording) Commander Dick Scobee said, *I won't lock mine; I might have to reach something.*

T minus 1 minute, 24 seconds: (From NASA transcript of on-board voice recording) Pilot Mike Smith said, *Ooh Kaaaay*.

T minus 1 minute, 4 seconds: (From NASA transcript of on-board voice recording) Mission Specialist Ellison Onizuka said, *Dick's thinking of somebody there*.

T minus 1 minute, 3 seconds: (From NASA transcript of on-board voice recording) Commander Dick Scobee said, *Uh huh.*

T minus 59 seconds: (From NASA transcript of on-board voice recording) Commander Dick Scobee said, *One minute downstairs*. NASA description of comment, "One minute till launch."

T minus 52 seconds: (From NASA transcript of on-board voice recording) Mission Specialist Judith Resnik said, *Cabin pressure is probably going to give us an alarm.* NASA description of comment, "Caution and warning alarm. Routine occurrence during pre-launch."

T minus 50 seconds: (From NASA transcript of on-board voice recording) Commander Dick Scobee said, *OK*.

T minus 47 seconds: (From NASA transcript of on-board voice recording) Commander Dick Scobee said, *OK there.*

T minus 43 seconds: (From NASA transcript of on-board voice recording) Pilot Mike Smith said, *Alarm looks good.* NASA description of comment, "Cabin pressure is acceptable."

T minus 42 seconds: (From NASA transcript of on-board voice recording) Commander Dick Scobee said, *OK*.

T minus 40 seconds: (From NASA transcript of on-board voice recording) Pilot Mike Smith said, *Ullage pressures are up*. NASA description of comment, "External tank ullage pressure." Macidull's comment: Inside the External Tank are the liquid hydrogen tank and the liquid oxygen tanks. At the tops of these tanks are vents where the pressure sensors are located. Low pressures could indicate a leak.

T minus 34 seconds: (From NASA transcript of on-board voice recording) Pilot Mike Smith said, *Right engine helium tank is just a little bit low*. NASA description of comment, "SSME supply helium pressure." Macidull's comment: The SSMEs (Space Shuttle Main Engines) are equipped with a 4,000 psi (pounds per square inch inside pressure) liquid helium bottle which supplies pneumatic pressure for various valve actuations, purging the

SSME fuel and oxygen lines, and for backup/emergency SSME shutdown.

T minus 32 seconds: (From NASA transcript of on-board voice recording) Commander Dick Scobee said, *it was yesterday too.*

T minus 31 seconds: (From NASA transcript of on-board voice recording) Pilot Mike Smith said, *OK.*

T minus 30 seconds: (From NASA transcript of on-board voice recording) Commander Dick Scobee said, *Thirty seconds down there.* NASA description of comment, "30 seconds till launch."

T minus 25 seconds: (From NASA transcript of on-board voice recording) Pilot Mike Smith said, *Remember the red button when you make roll call.* NASA description of comment, "Precautionary reminder for communications configuration."

T minus 23 seconds: (From NASA transcript of on-board voice recording) Pilot Mike Smith said, *I won't do that; thanks a lot.*

T minus 15 seconds: (From NASA transcript of on-board voice recording) Commander Dick Scobee said, *Fifteen.* NASA description of comment, "15 seconds til launch."

T minus 6.556 seconds: (From the Presidential Report) ***ME - 3 Ignition Command***. Macidull's comment: The first of the three main engines (MEs) is commanded to fire. The valves controlling the fuel (liquid hydrogen) and oxidizer (liquid oxygen) are opened, releasing these two chemicals into the sparks flying around below the engine nozzles. These three engines comprise only part of the launch propulsion for the Shuttle. The solid rocket motors (SRMs) provide the rest of the thrust before being jettisoned. The MEs are run on liquid fuel from the big brown external tank (ET). The ET actually comprises two tanks, one on top of the other—the top one contains liquid oxygen, the lower contains liquid hydrogen. The hydrogen is what burns. The oxygen enables the chemical reaction called burning. Three things are required for fire: fuel, an oxidizer or oxygen, and heat to start it. There is enough oxygen in the air for our automobiles to burn fuel (gasoline) or logs at our lower altitudes. At higher altitudes, the air becomes thinner. In space there is none. So, the Shuttle has to carry its own oxygen for the MEs. The SRMs, however, have a "solid" mixture of both fuel and oxygen mixed together. Because they are already mixed together, there is no "gas pedal" as with the "liquid" engines. And, when the SRMs are lit, they stay lit until they either burn out or are destroyed by radio signals from ground control. This is why the MEs are lit first. If there is a problem with them, they can be shut down. When the SRMs are lit, the astronauts are along for the ride until they burn out. And this is also why T minus 0.00 seconds is when the SRMs are lit. Before ignition, as seen many times on the news, there are sparks sprayed by spark generators below the ME exhaust nozzles. When the unburned hydrogen fuel and oxygen exits these nozzles, the sparks cause ignition. This negates extra weight for an on-board starting system. This "spark generator" can be functionally related to the spark plugs in your car. It causes the fire to start.

The Challenger launch propulsion thrust necessary to attain a space orbit is provided by these three MEs and the two solid-fueled SRMs. Like Roman candles, once ignited, the SRMs cannot be turned off. Also, the SRMs are contained in what is called the solid rocket booster (SRB). The SRBs include the SRMs plus vectored nozzles for direction control, and a nose cone with parachute for later recovery and reuse. As mentioned, once lit, the crew is captive for about a minute and fourteen-second ride. The biggest safety concern is that the solid rocket boosters cannot be discarded. If separated from the spacecraft while burning, they would accelerate past it and the blast from the exhaust would damage the orbiter, most likely beyond the possibility of recovery. Also, to give an indication of the inherent danger and requirement of quality control relative to the SRMs, if the propellant mix is not matched nearly perfectly in the left and right solid rocket motors, danger lurks. If one of them burns out as little as 1.7 seconds before the other, yaw control would be lost, and the entire Shuttle stack would spin out of control with no hope of recovery for either the machinery or the crew.

T minus 6.446 seconds: (From the Presidential Report) ***ME - 2 Ignition Command***. Macidull's comment: The number 2 liquid-fueled engine has been commanded to start. Liquid hydrogen and oxygen valves open into number 2-engine nozzle.

T minus 6.326 seconds: (From the Presidential Report) ***ME - 1 Ignition Command***. Macidull's comment: The number 1 liquid-fueled engine has been commanded to start. Liquid hydrogen and oxygen valves open into the number 1 engine nozzle.

T minus 0:000 seconds: (From the Presidential Report) ***SRM Ignition Command (T=0)***. Macidull's comment: This is designated "Time Zero" because, once the solid rocket motors are started, there is no turning back. During the past 6 seconds up until 0.00, when the launch clock turned from minus to plus, billions of computer-processed data bits completed measurements and tests. Had any of the tests proved abnormal, the launch would have been stopped until the problem was resolved. When Challenger was placed on the launch pad, all of its components were what is called a "stack." The stack consisted of the MEs, which were attached behind the Challenger's winged part, or orbiter, which held the crew and payload. The MEs were "on the centerline" of the Challenger. The solid rocket motors, however, were attached to the external fuel tank, which was in turn attached to the belly of the orbiter. The thrust from the MEs was therefore off-center from the solid rocket boosters, which provided the only attachment to the ground. As the main engines ignited, the thrust was along the centerline of the orbiter. The ME's thrust made the entire stack of hooked together components, the orbiter, ET, and SRBs (containing the SRMs), bend toward the solid rocket boosters and ET, and then back to vertical. This was called the "twang." The highest point on the stack is the top tip of the external tank. At twang on ME ignition, the bolts connecting the SRBs to the ground held the stack in place for a few seconds until ME thrust was computer verified. This twang caused the top tip of the ET to sway about a foot away from the vertical, also moving the orbiter and its MEs (which were not attached to the ground) away from the vertical. Without the stack pointing straight up, it would be risky to launch at this time. There were gantries and scaffolds to miss on pad 39B during the first few seconds of launch. Ten bolts held the entire stack to the ground. These were explosive bolts, five for each solid rocket booster. Once these SRBs and MEs obtained the necessary thrust for liftoff, and the twang returned the stack back to vertical, strain gauges on each of the bolts told the computer when the tension load on each was about equal. This meant also that the stack had twanged back to vertical, making it safe to clear the gantry. A switch was then computer-thrown to explode the bolts, sending

Challenger on its way. The system was also designed so that if any one bolt on either or both SRBs failed to explode, then the thrust from the engines and motors would break it or them in tension anyway and release the stack. This operation alone was an engineering marvel.

T minus 0 seconds: (From NASA transcript of on-board voice recording) Commander Dick Scobee said, *There they go guys.* NASA description of comment, "SSME Ignition." Mission Specialist Judith Resnik said, *All right.* Commander Dick Scobee said, *Three at a hundred.* NASA description of comment, "SSME thrust level at 100% for all three engines."

T plus 0.008 seconds: (From Presidential Report) ***Hold-down Post 2 PIC firing***. Macidull's comment: PIC (Pyrotechnics Initiator Controller) starts exploding the hold-down bolts, separating the stack from the earth. As the flight lifted off, the Challenger stack appeared to launch normally. On the ground, the Range Safety Office monitored the flight, watching for any problems. He would have to destroy the rockets and external tank if he saw the stack heading toward a populated area.

T plus 0.250 seconds: (From Presidential Report) ***First continuous vertical motion***. Macidull's comment: Challenger lifts off. **T plus 0.678 seconds:** (From Presidential Report) ***Confirmed smoke above field joint on RH SRM***. Macidull's comment: Smoke recorded on film seen coming out of the side of the right solid rocket motor.

T plus 0.836 seconds: (From Presidential Report) ***Eight puffs of smoke (from 0.836 thru 2.500 sec MET)***. Macidull's comment: Eight puffs of smoke recorded on film seen coming out of the side of the right solid rocket motor from T plus 0.836 through 2.500 seconds Mission Elapsed Time (MET) or from T=0 seconds.

T plus 1 second: (From NASA transcript of on-board voice recording) Pilot Mike Smith said, *Here we go.* NASA description of comment, "Vehicle motion."

T plus 2.733 seconds: (From Presidential Report) ***Last positive evidence of smoke above right aft SRB/ET attach ring***. Macidull's comment: Last smoke recorded on film where the right solid rocket booster (SRB) is attached to the external fuel tank (ET).

T plus 3.375 seconds: (From Presidential Report) ***Last positive visual indication of smoke.*** Macidull's comment: Last smoke from left SRM recorded on film.

T plus 4.339 seconds: (From Presidential Report) ***SSME 104% Command.*** Macidull's comment: Space Shuttle Main Engines (SSMEs) are throttled up to full power.

T plus 5.564 seconds: (From Presidential Report) ***RH SRM pressure 11.8 psi above nominal.*** Macidull's comment: The right hand solid rocket motor inside pressure is slightly above normal.

T plus 7 seconds: (From NASA transcript of on-board voice recording) Commander Dick Scobee said, *Houston, Challenger roll program.* NASA description of comment, "Initiation of roll program." Macidull's comment: The exhaust nozzles on both the SSMEs and the SRBs swivel to provide both directional and roll control for the entire stack. A computer program automatically commands these nozzles to point the stack in the required direction to enter the desired earth orbit in space.

T plus 7.724 seconds: (From Presidential Report) ***Roll maneuver initiated.*** Macidull's comment: As recorded on Shuttle-to-ground telemetry channels.

T plus 11 seconds: (From NASA transcript of on-board voice recording) Pilot Mike Smith said, *Go, you mother.*

T plus 14 seconds: (From NASA transcript of on-board voice recording) Mission Specialist Judith Resnik said, *LVLH.* NASA description of comment, "Reminder for cockpit switch configuration change." Macidull's comment: In spite of several requests during the investigation, NASA refused to provide the Presidential Commission with a copy of the Shuttle crew launch checklist. The meaning of "LVLH" therefore remains unknown outside NASA. However, speculation and some research of the acronym seems to indicate "Local vertical/local horizontal." But what this actually means remains hidden in the flight manuals not provided to the Presidential Commission after being requested.

T plus 15 seconds: (From NASA transcript of on-board voice recording) Mission Specialist Judith Resnik said, *(Expletive) hot.* Macidull's comment: Common aviation slang translation for "awesome" is "shit hot."

T plus 16 seconds: (From NASA transcript of on-board voice recording) Commander Dick Scobee said, *Ooohh-kay.*

T plus 19 seconds: (From NASA transcript of on-board voice recording) Pilot Mike Smith said, *Looks like we've got a lotta wind here today.* Macidull's comment: Cockpit thrust vectoring readings could indicate how much the SSME and SRB nozzles are moving to compensate for the wind and keep the stack on the desired flight path.

T plus 19.859 seconds: (From Presidential Report) ***SSME 94% Command.*** Macidull's comment: Space Shuttle Main Engines are throttled back to 94 percent power. This is a standard command. With a lighter weight because of the fuel burned off, the stack goes faster. Because of the thickness of the atmosphere and the high speed, the air pressure could injure the stack if a higher speed were attained at that altitude. They had to keep their speed down until they reached the thinner upper atmosphere. The only thrust sources able to be throttled back were the SSMEs.

T plus 20 seconds: (From NASA transcript of on-board voice recording) **Commander Dick Scobee said,** *Yeah.*

T plus 21.124 seconds: (From Presidential Report) ***Roll maneuver completed.*** Macidull's comment: The Shuttle is now pointed in the desired direction.

T plus 22 seconds: (From NASA transcript of on-board voice recording) Commander Dick Scobee said, *It's a little hard to see out my window here.*

T plus 28 seconds: (From NASA transcript of on-board voice recording) Pilot Mike Smith, *There's ten thousand feet and Mach point five.* NASA description of comment, "Altitude and velocity report." Macidull's comment: Altitude reported at ten thousand feet, and speed reported to be 0.5 Mach, or half the speed of sound.

T plus 35 seconds: (From NASA transcript of on-board voice recording) Commander Dick Scobee said, *Point nine.* NASA description of comment, "Velocity report 0.9 Mach." Macidull's comment: Speed reported to be 0.9 Mach, or 90% of the speed of sound.

T plus 35.379 seconds: (From Presidential Report) ***SSME 65% Command.*** Macidull's comment: Space Shuttle Main Engines were throttled further back to 65 percent power because of the lighter weight and increased speed due the fuel burned off in the still thick atmosphere. The SRMs could not be throttled and were still at full power at this time.

T plus 36.990 seconds: (From Presidential Report) ***Roll and Yaw Attitude Response to Wind (36.990 to 62.990 sec)***. Macidull's comment: Further investigation revealed that though the SSME and SRB nozzles swiveled nearly to their limits to compensate for some higher altitude wind shear, they operated normally as designed and within limits.

T plus 40 seconds: (From NASA transcript of on-board voice recording) Pilot Mike Smith said, *There's Mach one.* NASA description of comment, "Velocity report, 1.0 Mach." Macidull's comment: Speed reported to be Mach one, or at the speed of sound.

T plus 41 seconds: (From NASA transcript of on-board voice recording) Commander Dick Scobee said, *Going through nineteen thousand.* NASA description of comment, "Altitude report (19,000 ft.)"

T plus 43 seconds: (From NASA transcript of on-board voice recording) Commander Dick Scobee said, *OK, we're throttling down.* NASA description of comment, "Normal SSME thrust reduction during maximum dynamic pressure region." Macidull's comment: The SSMEs are being throttled further back to 64% power.

T plus 51.860 seconds: (From Presidential Report) ***SSME 104% Command.*** Macidull's comment: The atmosphere is now thin enough to throttle the main engines back up to full power.

T plus 57 seconds: (From NASA transcript of on-board voice recording) Commander Dick Scobee said, *Throttling up.*

T plus 58 seconds: (From NASA transcript of on-board voice recording) Pilot Mike Smith said, *Throttle up.*

T plus 58.788 seconds: (From Presidential Report) ***First evidence of flame on RH SRM***. Macidull's comment: Film records show flame for the first time coming out the side of the right-hand solid rocket motor.

T plus 59 seconds: (From NASA transcript of on-board voice recording) **Commander Dick Scobee said,** *Roger.*

T plus 59.000 seconds: (From Presidential Report) ***Reconstructed Max Q (720 psf)***. Macidull's comment: The combination of atmosphere thickness and speed has caused the maximum dynamic pressure on the stack of 720 pounds per square foot, (or 5 pounds per square inch.)

T plus 59.262 seconds: (From Presidential Report) ***Continuous well-defined plume on RH SRM***. Macidull's comment: Film evidence shows an obvious plume of flame coming out of the right-hand solid rocket motor.

T plus 59.753 seconds: (From Presidential Report) ***Flame from RH SRM in + Z direction (seen from south side of vehicle)***. Macidull's comment: Film from another camera shows an obvious plume of flame coming out of the right-hand solid rocket motor. "+ Z" is in the direction "down" relative to the Shuttle.

T plus 60 seconds: (From NASA transcript of on-board voice recording) Pilot Mike Smith said, *Feel that mother go. Woooo-hoooo.*

T plus 60.004 seconds: (From Presidential Report) ***SRM pressure divergence (RH vs. LH)***. Macidull's comment: Shuttle-to-ground telemetered data shows that the pressure inside the right solid rocket motor is lower than the pressure inside the left solid rocket motor, which further indicates a leak in the right motor.

T plus 60.248 seconds: (From Presidential Report) ***First evidence of SRB plume attaching to ET ring frame***. Macidull's comment: First film evidence that shows the flame from the right solid rocket motor burning into where it is attached to the lower part of the external tank.

T plus 62 seconds: (From NASA transcript of on-board voice recording) Pilot Mike Smith said, *Thirty-five thousand going through one point five.* Macidull's comment: Altitude reported passing thirty-five thousand feet with speed at one and a half times the speed of sound.

T plus 65 seconds: (From NASA transcript of on-board voice recording) Commander Dick Scobee said, *Rereading four eighty-six on mine.* NASA's description of comment, "Routine airspeed indicator check." Macidull's comment: Scobee reads airspeed of 486 knots.

T plus 67 seconds: (From NASA transcript of on-board voice recording) Pilot Mike Smith said, *Yep, that's what I've got too.*

T plus 70 seconds: (From NASA transcript of on-board voice recording) Commander Dick Scobee, *Roger, go at throttle up.*

T plus 73 seconds: (From NASA transcript of on-board voice recording) Pilot Mike Smith said, *Uh oh.* Macidull's comment:

This ended, according to NASA, all on-board voice recordings in existence.

During the one second between **T plus 72.500** and **T plus 73.500 seconds**, numerous telemetered readings were transmitted from Challenger to the ground. Based on this TM data, and other information addressed in this book, this is what physically happened according to Macidull:

When the O-rings failed, a flame like thousands of blow-torches, spewed white-hot gasses from the right solid rocket motor (SRM) toward the metal near the bottom of the brown external tank (ET). The entire round bottom of the hydrogen tank part of the ET was attached to the vertical sides of that tank by one long continuous circular weld. The flame from the SRM melted a small part of this weld. The pressure inside the hydrogen tank was 33 pounds per square inch (psi). The air pressure outside the tank in the atmosphere fell rapidly from about 14.7 psi at sea level to about 7 psi at 18,000 feet, then to about 3 psi at 46,000 feet at the time of the explosion. This outside pressure would have dropped to zero psi had the Shuttle reached the vacuum of space. When enough of the weld around the bottom of the liquid hydrogen tank melted from the flame shooting out of the side of the SRM, the pressure difference between the inside and outside of the hydrogen tank forced the entire weld to "unzip" (a term used relative to the action of a common zipper), causing the entire bottom of the hydrogen tank to come off. Had the hydrogen merely leaked from the hole in the weld, there would not have been an accident because not enough fuel could have leaked to place the Shuttle in critical jeopardy. If there had not been enough fuel left to get Challenger into orbit, after a few seconds when the solid rocket motors burned out and were automatically jettisoned, the crew could have easily jettisoned the external tank; and the Challenger could have glided back to Cape Kennedy or Africa for landing.

If leaking hydrogen from that tank had ignited from the flame of the leaking SRM, the Shuttle still would not have been in a critical situation. There is little oxygen at 46,000 feet to sustain burning. And even if leaking hydrogen did burn where the SRM flame was impinging on the hydrogen tank, there was no physical means there for the flame to propagate into the hydrogen tank, where there was no oxygen anyway. But because the bottom of the hydrogen tank unzipped, it was all over, as related below.

The three MEs produce about 375,000 pounds maximum thrust each. The two solid rocket motors produce about 2,800,000 pounds of thrust each. These five thrust sources are attached to mounts that can take the strain required to propel the entire stack. The brown external tank, consisting of a hydrogen tank and an oxygen tank, is not mounted to provide any thrust, either to the Shuttle or to itself. The 27.5-foot diameter hydrogen tank (the bottom tank inside the ET) had about thirty PSI of pressure difference at 46,000 feet, pushing against its inside walls according to recorded telemetered data sent to the ground.

As the unzipping of the weld occurred, pressure forced the hydrogen tank contents out of the bottom in an instant. This equates to 2,565,902 pounds of instantaneous thrust at a 30-psi pressure differential. Since the hydrogen tank was not mounted with the strength to provide any thrust to the Shuttle, it instead thrust itself straight up and into the oxygen tank (the top tank inside the ET). The now-ruptured oxygen tank then spilled liquid oxygen that immediately vaporized and, because of the airstream, was blown back. It blew back into both hydrogen and flame. Now there was an ample amount of all three ingredients for an "explosion"—fuel (hydrogen), oxidizer (pure oxygen), and heat (solid rocket motor flame leak).

However, the explosion just changed the flight vector of the Challenger while causing some minor damage. Air loads (dynamic pressure or Q) at twice the speed of sound at that altitude then ripped Challenger and the entire stack apart. It is interesting to note that the exhaust from a hydrogen/oxygen engine is water, H_2O. The place of this explosion has often been identified as a big area of mostly white smoke. This was not smoke; rather, it was water vapor, a man-made cloud from the burning of hydrogen and oxygen, which results in H_2O.

At **T plus 72.564 seconds** the Challenger stack began to come apart. Also between **T plus 73.482** and **T plus 73.523 seconds,** the onboard computer, because of over-temperature detected by internal sensors, automatically shut down all three main engines. The oxygen tank was farther away from the main engines (MEs) than the hydrogen tank. With both tanks gone, there was more oxygen left in that line to the engines than what was left in the hydrogen line. A proportional increase in oxygen, with the sustained ignition inside the engines, caused a higher temperature and began burning the engines themselves from the inside out. Sensors recognized this abrupt increase in temperature and decrease in hydrogen, and the on-board computer program shut the engines down. Later wreckage inspection showed that the main engines had suffered internal burn damage before being shut down.

At **T plus 74.587 seconds** the Shuttle Challenger stack exploded. In a scientific sense, however, there was no explosion. There was no stoichiometric balance for the perfect ratio of hydrogen and oxygen. It was a "slow burn" in explosion parlance. For the purpose of this history, however, the word explosion is used.

During this explosion, the gases impacted the vulnerable orbiter at supersonic speed and destroyed its vector of flight, in a still dangerous atmosphere of air, at the speed it was going. The Challenger's wings could not take more than one "G." With the abrupt change in direction, its wings absorbed the new lift angle and broke off. The air molecules at that speed and altitude shredded most of the rest of Challenger. Engines sputtered, and the solid rocket boosters became disconnected from the external tank but continued to thrust, out of control, accelerating ahead and away from Challenger. When the stack came apart, Challenger was going nearly straight up at almost twice the speed of sound (Mach 1.92) at 46,000 feet. All that kinetic energy didn't stop Challenger's pieces at that place in the sky. The crew compartment, with the 7 astronauts still strapped in their seats, exited the explosion and continued up. The compartment was slowed by air friction and gravity and came nearly to a stop at about 63,000 feet. Then gravity pulled it back toward earth. Some of the crew, according to hard physical evidence obtained after the Presidential report was published, were breathing for about two minutes or longer during this time after the explosion. This suggests that all of the crew were breathing during this time and all personal crew equipment had been obtained during the investigation for evaluation. Although there was no propulsion source of any type for the truncated crew compartment, the compartment sped up from the Earth's gravitational pull. However, without any propulsion, its terminal velocity in the ever-thickening air of lower altitudes limited its speed to about 200 miles per hour as it approached the ocean. On impact, all were killed.

End of this annotated flight sequence.

An extended, non-verified crew voice transcript of the onboard tape from approximately the time of the explosion to water impact (T + 1:15 to T + 3:15) has been published in Weekly World News and has circulated on Usenet and several web sites. It can be found under the search topic, "Challenger's Final Minutes." NASA denies the existence of this part of this tape. Though this alleged extension of the transcript appears reasonable to this author because of a reference to an "air pack" (PEAP Bottle stands for Personal Emergency Air Package Bottle which each crewmember had), it could have been written by someone with specific technical knowledge of the shuttle, and cannot be considered germane to the purposes of this history. Also, if it were factual, it would only reinforce what has already been determined based on other hard evidence.

APPENDIX III

A BRIEF HISTORY OF SPACE FLIGHT, THE POLITICAL EVOLUTION OF THE SPACE SHUTTLE, AND THE GAMBLING AWAY OF NATIONAL SECURITY

By Blattner

On October 4, 1957, the Soviet Union successfully launched Sputnik I, the world's first man-made satellite. Congress, alarmed, rushed the National Aeronautics and Space Administration, NASA, into creation. When President Dwight D. Eisenhower signed the National Aeronautics and Space Act into law on July 29, 1958, he created an organization to "...plan, direct, and conduct aeronautical and space activities," to "...arrange for participation by the scientific community in planning scientific measurements and observations," and to "...provide for the widest practicable and appropriate dissemination of information concerning its activities and the results thereof."

No matter what the enabling legislation creating the agency said, NASA was a creature of the so-called space race. Congress actually appropriated more money than NASA needed. And NASA spent nearly all of this money on space, letting aeronautics and, thereby, airline safety roll into the backwaters of neglect for more than thirty years, long after the space race was conclusively won.

Then came the orbital flight of a Russian, Yuri Gagarin on April 12, 1961. That flight put tremendous pressure on NASA to get the United States space program moving. Finally on May 5, 1961, after a one-day delay because of technical problems, Alan Shepard rocketed off from Cape Canaveral to become the first U.S. man in space aboard Freedom 7, a one-man Mercury spacecraft boosted by an Army Redstone rocket.

As the second American in space, John Glenn would orbit the earth three times on May 15, 1963. Yet Glenn's flight was also delayed, this time for more than a month because of problems. In retrospect, the lengthy delays in the countdown were among NASA's finest hours. No matter how long it took, or how difficult the problem, an astronaut was not allowed to fly until all the spacecraft and rocket systems were as safe as humanly possible.

By the 1970s, the evolutionary Apollo program had provided some engineering moments of tremendous import and respect, especially the legendary Apollo 13 mission. With that craft in deep space, a malfunction cropped up. NASA mission controllers thought it was the electrical system that failed, causing a possible disconnect between fuel cells. Millions of people watched around the world as the drama unfolded on their television sets. The rescue operation, through flexible, brilliant, and innovative engineering skills on the ground, brought Commander James Lovell, command module pilot Thomas Mattingly, and lunar module pilot Fred Haise safely back to earth.

How did NASA, the government agency responsible for manned space exploration since 1959, arrive at a time where safety concerns, no matter how diverse, became less critical to a launch decision? Perhaps this question is best answered as follows.

In 1968, NASA was similar to other Washington agencies such as the Departments of Commerce, Defense, or Education. NASA employed about 30,000 hard-working, dedicated personnel. According to Macidull, numerous NASA officials felt the pain of the Challenger loss more than most Americans. But at the time of Challenger, NASA was undergoing a major cultural upheaval.

Early in 1985, NASA Administrator James Beggs was indicted for fraud. He was later exonerated of the charge and vacated NASA in November 1985, prior to the Challenger launch. It was a tumultuous time within NASA's culture, with unsettling consequences within the community of scientists, engineers and managers. "It was an unfortunate turn of events," says Beggs, "but I do not believe anything that was going on at headquarters had anything to do with the Challenger accident." Still, Dr. William Graham, NASA's Deputy Administrator, was then running the agency. The transition and controversy surrounding the Beggs investigation was unsettling and somewhat demoralizing to numerous personnel, if not at Cape Canaveral, then certainly to the dedicated scientists and engineers who depended upon good leadership for money from Congress.

But if this had nothing to do with the launch of Challenger, why bring it up here? We must do it because NASA covered up facts of the Challenger crash and how the astronauts died, and because NASA secretly cut deals with the families of the astronauts without telling the American public. Because, even today the Freedom of Information Act cannot be used at NASA without the agency crossing out page after page of legal and clandestine arrangements with Thiokol after the crash.

There were several other factors working against NASA at the time of the accident. But none was more pressing than NASA's budget. In the late 1960s NASA's big push in Congress was for funding necessary for a manned mission to Mars. This would not be an easy task.

Space had become a passé subject for the American people. They had seen man walk on the moon but were now more intent on considering the impact of the war in Vietnam. In the 1960s NASA was actually competing with wartime expenditures for their budget.

NASA's problem was summed up in a July 1969, Gallup Poll. The poll found 39 percent of 1,517 persons polled nationally favored attempts to land a man on Mars, while 53 percent opposed such a plan. This was largely along age lines. Of 21- to 29-year-olds, 54 percent favored the project and 41 percent opposed. But 60 percent of those over 50 opposed. Perhaps the thumbs-down reaction to a Mars flight was to be expected, given the projected cost of such a program, which was estimated at $54.1 billion to $78.2 billion during the 1970-1980 decade.

Says Beggs, "The chief proponent for going to Mars was Werner von Braun. At that time he was in charge of NASA's planning and was pushing hard for a trip to Mars. As they talked about it they realized that in order to go to Mars, you were talking about a 200-day trip, at best. There was much that needed to be researched and technologically developed. We had worked on a nuclear rocket, Project Nerva, but that project had been terminated. Still, the most serious impediment to the Mars trip was life support. We didn't have the technology for artificial gravity, for example. We estimated that the first thing we needed was a space station.

"So we went to President Nixon with a proposal to go to Mars. In order to do this, we needed to solve a number of problems. But what we needed in the near term was a space station and a space shuttle-a transportation system that would get us back and forth reliably, be efficient, and fit within our budget. We wanted to achieve a low orbit for less than $1,000 per pound. One has a feeling of *deja vu* these days since they are still talking about $1,000 per pound."

In July 1969, Air Force One winged westward across the Pacific toward the splashdown target of Apollo 11. The four men in that plane were President Richard M. Nixon, Secretary of State William P. Rogers, National Security Adviser Henry A. Kissinger, and then-NASA Administrator Thomas O. Paine. All aboard the airplane knew that the Apollo program "was a watershed, making the first lunar landing and those who would follow, the end of an initial phase of space exploration," according to a NASA document.

To all who knew the space program, by 1969, the innocent era of Mercury, Gemini, and Apollo had passed. It had been a storybook beginning to one of man's greatest adventures. But what would follow? How would Congress greet large new budgetary expenditures required for new space programs? How could the White House rally the American public to support such expenditure?

By September 1969, two months after the meeting in Air Force One over the Pacific, a Space Task Group chaired by the Vice President created a report, *Directions for the Future.* It told that the nation had backed away from an early manned landing on Mars. "That didn't bother von Braun in the least. He expected it. But the budget situation turned critical and caught many off guard," said Beggs.

In the report, the group had developed three long-range plans: a $8-$10 billion-per-year program involving a manned Mars expedition, a space station in lunar orbit and a 50-person Earth-orbiting station serviced by a reusable ferry or space Shuttle. An intermediate program, costing less than $8 billion annually, that would include the Mars mission; and finally a relatively modest $4-$5.7 billion a year program that would embrace an Earth-orbiting space station and the Space Shuttle as its link to earth. In March 1970, President Nixon made clear that he favored a continuing active space program. The President then opted for the Shuttle-tended space base as a long-range goal; but he deferred going ahead with the space station, pending development of the Shuttle vehicle. So that was it. The Space Shuttle, earlier considered simply a transport element of a broad, multi-objective space plan, was now the main concern for NASA's future.

Still, six months would pass before President Nixon recognized the task group's findings. By that time, however, Congress, during the appropriation process, had shaped the future for NASA's programs by restricting the agency's budget even further. The Office of Management and Budget (OMB) actually set a spending limit from the beginning of the program in 1971- dollars at $1.1 billion. Even so, cost over-runs in developing the Shuttle for two decades forced the cost of each Shuttle to exceed original estimates by 5 percent.

The 1970s were difficult for NASA. The decade began with the agency taking its budget to Congress and seeing it slashed from more than $4 billion to $3.3 billion, down more than $500 million from the previous fiscal year. Then-Administrator, Thomas Paine told the House Committee on Science and Astronautics that NASA had to make some "hard choices" to stay within that budget while still making progress: suspending production of Saturn V launch vehicles and spacecraft assigned to those vehicles; mothballing the Mississippi Test Facility and reducing the work force at the Michoud Assembly Facility where Saturn Vs were assembled; closing the new Electronics Research Center at Cambridge, Massachusetts; and canceling Apollo 20. It was a marked difference from the previous decade, when Congress threw money at NASA without any reservations, in order to put a man in space. But nonetheless, by 1976 the Space Transportation System (STS), later to be called the Space Shuttle, was well in focus. Early estimates espoused ambitious requirements, such as an eventual launch rate of 75 Shuttles per year from both Cape Canaveral in Florida and Vandenberg AFB in California. This was later toned down to 60 Shuttle flights annually from seven Shuttles, three launch pads, and two orbiter-processing facilities. In spite of a more conservative budget from Congress, NASA's plan also included a space station and space tug that could work on and move geo-synchronous satellites in earth orbit.

Such a concept was a draft blueprint, and NASA wanted to replicate a fast launch turn-around rate with as many Shuttle flights as possible every 12 months. Even by 1979, Robert A. Frosch, NASA's administrator from June 21, 1977, to January 20, 1981, was still making public promises that the Shuttle would be launched into space "every Monday morning like clockwork." The goal was a proven Shuttle program that was fast, efficient space transportation deserving a larger congressional investment.

Was NASA promising too much to Congress during this time? "Yes," says Beggs. "You don't create and sell a program in this country to Congress unless you oversell it. That's true not only of space programs. I was involved at the birth of Medicare and Medicaid, and those programs were oversold on the basis that they would cost no more than $5 million a year. They are billion-dollar programs today. In short, even the military has learned to oversell programs to make them palatable to Congress and the public. So yes, we oversold the space program. Ultimately we were able to fulfill the promise. But in truth, we oversold the program."

By 1980, the United States' leadership and preeminence in space were seriously threatened. The Soviet Union had established an essentially permanent manned presence in space with its MIR space station and was using the space station to meet economic, military, and foreign policy goals. Japan was broadcasting directly from space to individual homes and businesses, and France was moving ahead of the U.S. in preparing to reap the economic benefits of satellite resource observation.

U.S. commercial enterprises were turning to France to launch their satellites. So a Congressional investment was what NASA meant when talking about the Shuttle program to Congress. The idea was to use public money to bankroll further manned space programs. For example, NASA told Congress that the Shuttle would be a cargo carrier delivering cargo at less than $1,000 a pound. By the time of the Challenger accident, that cost was actually over $6,000 a pound. By the 1980s the Shuttle carried 41 percent of all the bulk cargo, including satellites, into space.

By the end of the decade, NASA had a well-heeled customer, the U.S. Air Force. Of course this arrangement came back to haunt the military after the Challenger accident. Between 1986 and 1988 the United States had no capability to launch military satellites because, in the arrangement with the Pentagon, the Air Force decommissioned its Atlas and Titan space boosters. This would turn out to be a major problem.

By 1984, NASA planners were already figuring out ways to use launches to draw public support to the agency. Florida Congressman Bill Nelson was recruited to fly aboard Columbia on mission 61C, the 24th Shuttle launch. Unfortunately for NASA, Nelson did not run again for Congress in the November elections; and by the time the long-delayed Columbia flew, two weeks prior to Challenger on January 12, 1986, Nelson was no longer a Congressman.

For those older and dedicated space officials, the always publicity-seeking NASA had become a circus. The Challenger's crew consisted of seven Americans who were chosen to "personify the nation's heterogeneous mix of gender, race, ethnicity, and age. NASA's public relations staff focused media attention on Christa McAuliffe, the New Hampshire high school social studies teacher. The media embraced the idea, and overnight McAuliffe became a household name. The object seemed to be to show the public how ordinary and safe space flight had become. McAuliffe was assigned the title of Payload Specialist One. She was a major part of the Challenger attraction.

Finally, by the time Challenger was launched, NASA had lost some of its institutional memory. Gone were men such as Christopher C. Craft, Jr., past Director of Marshall Space Center, and Robert R. Gilruth, once NASA's Director of Personnel Development. Both Craft and Gilruth were original members of the NASA Space Task Group established in 1958 to manage Project Mercury. They literally wrote the book on how to launch man into space. But by 1986 there was a new generation of baby boomers who were highly credentialed managers, responsible for the human life of Shuttle crews.

These NASA managers rolled up an impressive record of 24 straight launch successes. But were they doing everything correctly in launching these missions? Had people such as George Hardy, Arnold Reinartz and Lawrence Mulloy, knowledgeable and experienced engineers, understood this trend, and listened to what Thiokol was saying about the solid rocket motors, the Challenger launch would have been called off until either the problem was fixed, or a decision would have been made to launch on a warmer day.

Of course, there was so much false testimony before the Commission surrounding what happened during those final hours that we can only piece together what really happened. In fact, those in the trenches were doing their best to thwart anyone who was trying to expose negative information about the agency. So it is pertinent to this story to have some first-hand experience with NASA after the crash.

Still, another event of importance in this story also happened on that day, but there was no photo opportunity. The National Security Decision Directive (Number 42) on National Space Policy was released in Washington. If there was ambiguity in the NASA-enabling legislation that stated that the space agency would develop programs in pursuit of "peaceful purposes for the benefit of all mankind," the new National Space Policy erased them. That document says:

"The basic goals of United States space policy are to: (a) strengthen the security of the United States; (b) maintain United States space leadership; (c) obtain economic and scientific benefits through the exploitation of space; (d) expand United States private-sector investment and involvement in civil space and space-related activities; (e) promote international cooperative activities that are in the national interest; and (f) cooperate with other nations in maintaining the freedom of space for all activities that enhance the security and welfare of mankind."

Security is mentioned twice in this statement, including the first item of the policy. Since only one manned space vehicle was flying, the Shuttle, it seemed inherent in the security policy of the nation. Further, Section II of that document, the Space Transportation System, specifically stated:

"The United States is fully committed to maintaining world leadership in space transportation with an STS capacity sufficient to meet appropriate national needs. The STS program requires sustained commitments by all affected departments and agencies. The United States will continue to develop STS through the National Aeronautics and Space Administration (NASA) in cooperation with the Department of Defense (DOD). Enhancements of STS operational capability, upper stages, and efficient methods of deploying and retrieving payloads should be pursued as national requirements are defined."

This policy formalized what had been taking place behind the scenes by the military in the Shuttle program since its inception. The full military use of the program for national defense was now official. And with the language in place, the Air Force was now free to pursue use of the Shuttle not only as a launch pad for military spy satellites but also for undisclosed purposes as well.

Gambling Away National Security

Apollo Spacecraft Program Office (ASPO) Manager, George M. Low, was a NASA team leader on the NASA transition team in 1980. In a report of his activities he defined the space agency's mission as follows:

"The Act [NASA's enabling language] declared that it is the policy of the United States that activities in space be developed to peaceful purposes for the benefit of all mankind, and that these activities (except those primarily associated with the defense of the United States) should be the responsibility of a civilian agency. This agency—NASA—was chartered to carry out significant programs in aeronautics, space science, space technology and applications, and manned space flight.[1]"

Given the above definition of NASA's space policy, what was the military doing in a peaceful space program? What did this mean to Challenger? To answer these and other relevant questions concerning the military and NASA, it is appropriate to look at the nation's security policy.

Up until the Shuttle program, the U.S. Constitution defined this policy as: ..."To provide for the common defense, promote the general Welfare, and secure the Blessings of Liberty to ourselves and our Posterity...[2]

According to the founding fathers, such basic requirements were non-negotiable. They had been constant throughout the nation's history. They became negotiable, however, in dealings between NASA and its major customer, the U.S. Air Force.

The Shuttle program was not the first involvement NASA had with the military. The first manned flight was boosted on a sub-orbital flight with an Army ICBM—the Redstone. The Atlas and Titan II rockets served as Mercury and Gemini spacecraft boosters.

There was also a proposal based on NASA's peaceful Gemini program, for an Air Force "Blue Gemini"—tag named for an Air Force manned space flight program to develop rendezvous, docking, and transfer for military purposes, using Gemini-type spacecraft. The proposal first surfaced in February 1962, during congressional hearings on the defense budget. Blue Gemini, it appeared, was part of a far-ranging Air Force space plan for the development of military space technology over the next decade.

The concept became more tangible when the Air Force Space Systems Division began working on plans to use Gemini hardware as the first step in a new Air Force man-in-space program called MODS (Manned Orbital Development System). This was a military space station with Gemini spacecraft used as a ferry vehicle. The following year, the term Blue Gemini was used as part of a more specific proposal to fly six Gemini missions with Air Force pilots in a preliminary orientation and training phase of MODS.[3]

Blue Gemini, however, never made it off of the drawing boards. It did not have the broad based political support necessary to fund such projects. It did not even have good support within the Pentagon. Specifically, the Air Force Chief of Staff, Curtis E.

LeMay feared NASA entanglement would endanger funding for other military programs. The failure of Blue Gemini further indicated the absence of any clear-cut military need for manned operations in space.

Yet, by mid 1970, NASA discovered it could not function without the military. At this time, NASA was pitching Congress on a way to supplement budget expense through the use of the Shuttle for paying customers.

The agency lobbied both Congress and the President saying the Shuttle would cost about one-third the cost of expendable launch vehicles with comparable payload capacity. The Shuttle would be ready for a preflight only after seven days of maintenance and checkout, resulting in a flight rate of about 55 launches per year for the entire Shuttle fleet.

A final point of the NASA presentation was that: "all five Shuttles would be built to have performance capability sufficient to fly the largest U.S. payloads into orbit." In other words, NASA claimed the Shuttle would cost one-third as much to put a satellite into orbit as a Titan would. And, the Shuttle would be turned around in seven days and place in orbit even the heaviest satellite.

There were those in the Defense Department who doubted NASA's claims about the Shuttle for this purpose, yet there wasn't sufficient documentation to support their charges, according to E. C. "Pete" Aldridge, Jr., then Under Secretary of the Air Force, who subsequently became Air Force Secretary in 1984. [4]

Aldridge says that "NASA grossly exaggerated the Shuttles capabilities in selling the concept to Congress. Furthermore, NASA told Congress that it could help pay for the program by getting paying customers to use the Shuttle to launch civilian and military satellites. The agency was talking about the Air Force as its major customer."

To help sell its idea to Congress, Aldridge said that NASA claimed the Shuttle would cost about one-third the cost of expendable launch vehicles with comparable payload capacity. The Shuttle would be ready for a re-flight only after seven days of maintenance and checkout, resulting in a flight rate of about 55 launches per year for the entire Shuttle fleet. A final point of the NASA presentation said that, "all five Shuttles would be built to have performance capability sufficient to fly the largest U.S. payloads," according to Aldridge.

According to Aldridge, NASA said the Shuttle would cost one-third as much to put a satellite into orbit as a Titan-boosted satellite. The Shuttle would be turned around in seven days, and orbit even the heaviest weight satellite He said that these promises are part of the public record.

Aldridge remembers this time and the arguments of those within the Defense Department and Air Force who did not favor the policy. "Some in the Pentagon and Air Force opposed the policy; yet at that time, they had no hard data on which to argue their point. They said things like, 'we don't think this thing is going to be cost effective, we don't think it's going to have the turn-around time, we don't think it's going to have the performance as advertised.' Still, there was simply no proof to back up these arguments. In fact, if the figures NASA was touting on performance and cost of the orbiter were right, it would allow the Air Force to reduce its launch costs, and then put that money into satellite programs. Those in favor of the concept thought they could make their programs more robust with the savings from the Shuttle program," he said.

Even in 1963, the relationship between NASA and the Air Force was in full force. While astronaut candidates had originally been chosen from the military ranks, they were plugged into a civilian space program. Aldridge, for example, was not only a political official in the Air Force; he was also a pilot and astronaut candidate for the first Shuttle military mission from Vandenberg AFB. It was a position he continued in until after the Challenger accident, when he returned to Washington and his job in the Pentagon. Today Aldridge is the President and CEO of the Aerospace Corporation, in El Segundo, California. He spoke to us from the company's Washington corporate offices in Arlington, Virginia.

In viewing that period of history, NASA worked diligently in lobbying Congress. Influential congressional members were willing to go along with the NASA/Air Force launch decision. They apparently succeeded because congressional leaders told the Pentagon there would be no future funding for Defense Department satellites unless they were designed to fit in the Shuttle bay.

By the time the first Shuttle flew – Columbia (STS-1), April 12, 1981, the administration had changed and Ronald Reagan was President. By June 27, 1982, Columbia again flew on the fourth launch (STS-4) with the first Defense Department satellite. The Shuttle was now fully operational. It was a major triumph for NASA and the new President, who attended the landing at Edwards Air Force Base, California, on July 4, 1982, to mark the occasion.

Another event was happening on that day, which received no photo opportunity. The National Security Decision Directive (Number 42) on National Space Policy [5] was released in Washington, DC.

If there was ambiguity in the NASA enabling legislation that stated that the space agency would develop programs in pursuit of "peaceful purposes for the benefit of all mankind," the new National Space Policy erased them. Here was the language stated within that document:

"The basic goals of United States space policy are to: (a) strengthen the security of the United States; (b) maintain United States space leadership; (c) obtain economic and scientific benefits through the exploitation of space; (d) expand United States private-sector investment and involvement in civil space and space-related activities; (e) promote international cooperative activities that are in the national interest; and (f) cooperate with other nations in maintaining the freedom of space for all activities that enhance the security and welfare of mankind."

Security is mentioned in that statement two times, including as the first item of the policy. Since there was only one manned space vehicle flying, the Space Shuttle, it seemed inherent in the security policy of the nation. Furthermore, in Section II of that document, the Space Transportation System, it was specifically stated that:

"The United States is fully committed to maintaining world leadership in space transportation with an STS capacity sufficient to meet appropriate national needs. The STS program requires sustained commitments by all affected departments and agencies. The United States will continue to develop the STS through the National Aeronautics and Space Administration (NASA) in cooperation with the Department of Defense (DOD). Enhancements of STS operational capability, upper stages, and efficient methods of deploying and retrieving payloads should be pursued as national requirements are defined."

The policy formalized what had been taking place behind the scene by the military in the Shuttle program since its inception. The full military use of the program for national defense was now authorized. And, with the language in place, the Air Force was now free to pursue use of the Shuttle not only as a launch pad for military spy satellites, but for undisclosed purposes as well.

Still, there were other problems the Air Force had not anticipated. For one thing, the turn-around time of seven days between Shuttle lunches NASA had promised was in reality 60-days. For another, the launch cost was at least equal to the purchase of a Titan rocket. Then finally, according to Aldridge, Columbia could not lift the payloads that the Defense Department needed.

Most of all, NASA was having difficulty keeping up the schedule that the agency had promised. Said Aldridge, "The real costs to launch the Shuttle were becoming apparent and low launch rates forced the Air Force into larger than anticipated overhead costs. These had to be spread over fewer flights.

Then also, NASA had said that all five Shuttles would be almost identical. In fact each was substantially below the design specifications. NASA had promised that the orbiter would deliver 65,000 pounds of payload to low earth-orbit from Cape Canaveral.

The agency claimed it could place 32,000 pounds from polar orbit from Vandenberg AFB. Still, the final Shuttle capabilities were nearly 20 percent short of these goals. "The Air Force had built Vandenberg for the Shuttle flights. NASA didn't pay for that; the Air Force did. We built the inertial upper stage of the Titan that NASA used. The Air Force paid for that. As part of the partnership, the Air Force was required to pay for the expendable cost of each flight rather than the pro-rated share of the flight spread over the entire cost. Our expendable cost was the refurbishment of the solid rocket motors, the expense of throwing away the external tank and any unique software changes. We also paid for the fuel expended on each flight. That came out to be about $85 million per flight," Aldridge said.

By the fourth Shuttle mission, however, things that had been previously accounted for by NASA budget for Shuttle flights began to find their way into the Air Force expenditure. "This included even so-called minor items. For example, the payload tie-down bolts in the Shuttle bay; NASA was going to pay for those. Pretty soon NASA – in order to keep the price of the flight down – began forcing the payload customer to pay for even the tie-down bolts. This created a sour taste for the Air Force because we had not budgeted for such expenditures. All of a sudden we had to go pay for such items because NASA insisted that the Air Force pay the bill," Aldridge said. Such shortfalls became a major headache for the military. The Defense Department had taken NASA at its word and created satellites based on projected performance standards. When the Shuttle couldn't perform accordingly, NASA was forced to upgrade a filament wound case for the Shuttle's solid rocket motor and tweak the main Shuttle engines to perform at the higher thrust level to support the weights of the military satellites. All of this cost money, and NASA continually found additional ways to charge customers.

In 1983, Aldridge, as Under Secretary of the Air Force, was in charge of the Air Force military satellite program, said, "I wanted to keep the expendable launch vehicles going for a few more years until the Shuttle had demonstrated performance capabilities far beyond what we had seen so far. The Defense Department needed about eight Shuttle missions a year. NASA was proposing to fly 24 flights a year and we would get a third of them and that was okay. Yet there we were only seeing 12 flights a year. They might as well paint the Shuttle blue, put US Air Force on its side, and send it to Pentagon."

Meanwhile, the Shuttle was now creating doubters under real launch conditions.

Instead of five orbiters only four were built, and there were other differences as stated from NASA projections. Such doubts took substance in December 1983, when the Air Force presented a classified briefing to the Joint Chiefs of Staff and Defense Secretary concerning a "Net Assessment of U.S. and Soviet Space Capabilities."

In this presentation, there was a comparison of the Soviet Union and American space launch capabilities. This included concerns about a Shuttle-only launch capability for U.S. defense satellites. The briefing prompted Secretary of Defense, Caspar Weinberg to write a letter to the President. In this January 23,

1984, letter, Weinberg outlined the intelligence community's reluctance to go forward with only the NASA/Air Force launch schedule. "The country made a serious mistake in adopting this policy," he said.[6] Weinberg recommended that a limited number of earth launch vehicles be used to support Shuttle flights in launching Defense Department satellites.

On February 23, within days of the Weinberg letter, the Air Force found itself testifying before a hostile House Subcommittee on Space and Science Applications. The Air Force outlined its ideas and said that the earth launched vehicles, or ELV's would complement the Shuttle program. Under the proposal, the Air Force would continue to fly eight to ten missions annually, and use two ELV flights annually to hedge against technical risk and reasonable schedule availability. The Air Force said it was important to start this program immediately since by the end of 1984, the Atlas booster production would begin shutting down, causing the production base to erode. Judging from the hostile reception the Air Force received during the Committee proceedings, the congressmen were seemingly unimpressed, according to Aldridge.

Aldridge said that in May 1984, a memorandum arrived at the Defense Department from the NASA Administrator to the Secretary of Defense. In that memorandum NASA argued that the ELV plan was not necessary. Said the memo, "...a Shuttle derived launch vehicle would have growth potential to meet future space launch requirements of the nation, not only the Air Force."

Aldridge says that the Air Force had been put on the defensive and found itself in a very difficult and unpopular position. He said that by the summer of 1984, the Air Force's approach to use two ELV's annually to complement the Shuttle launches was fully developed. It called on contractors to pay for the development cost for a "commercial" ELV capable of carrying Shuttle-sized payloads. Contractors would be permitted to amortize costs over the number of lunch vehicles sold during the 1988-92 period. Government for its part would pay for the total launch one year in advance as it had done for the Shuttle. There was a major problem with this idea, however. Congress and the Office of Management and Budget (OMB), however, did not support the concept.

In June 1984, the Air Force had to pursue its conventional development and procurement program. This meant that the required funds for the ELV program would come from the Defense Department's Five Year Defense Program. Congress did approve the request for $5 million to design the new complementary EVL, known then as the CEVL. Congress also requested a joint study of the requirements, options and costs of the CEVL concept, and that the study be coordinated with NASA and the Defense Department and approved by the President. Part of the Congressional stipulation was that the President's approval had to be by January 15, 1985.

The infighting between NASA and the Air Force became nearly terminal over this proposal. NASA wanted Congress to demand that any unmanned vehicle had enough capability to lift 100,000 to 150,000 pounds into low earth-orbit. Obviously the system would be based on Shuttle -derived components and developed by NASA. The agency also wanted to compete with industry for the new system to meet Air Force requirements.

The Air Force countered that there was no requirement for such a system. Aldridge specifically said, "There was no known satellite launch requirement for a new, expensive space booster development based on undefined satellite launch requirements."

The battle went back and forth until Congress finally pressured the Air Force to accept NASA proposal, after eliminating the competition with industry. For its part, the Air Force decided that if it had to have this shotgun marriage, it would select a winner from an industrial competition and compare it to the NASA entry.

NASA continued its fight, and then in September 1984, a National Research Council study requested by Congress was released supporting the Air Force position. In making its recommendation, the study noted that it supported the plan to launch smaller payloads such as weather satellites on converted ICBMs such as the Titan III.

Said Aldridge, "When the bids came in, I had three proposals, one from General Dynamics, one from Martin Marietta, and one from NASA. Here I am trying to decide which proposal to take, and there was a government agency competing against industry. Was that legal?" A short time later, the Air Force announced that indeed a modified Titan III, Titan 34D7, was the winner of the industrial competition. It met the requirement to place 10,000 pounds into geo-synchronous orbit from Cape Canaveral at the lowest cost. It also could meet the Air Force flight-date requirement of October 1988. Said Aldridge, "The competition was literally a farce. If it wasn't, it should have been an embarrassment to NASA's highly competent engineering team." Losing the competition didn't stop NASA from making trouble for the Air Force. The agency requested that a yearlong study be completed to allow time for a proper solution to be determined. " It was a thinly veiled proposal, however, and without any financial commitment during the rest of the year, the ELV production lines would cease to exist. Any attempt to reopen them at a later date would be impossible because of cost," said Aldridge.

There were those within NASA who wanted to find a compromise and stop the inter-agency feud with the Air Force. NASA/DOD mission 51-C, Discovery, on January 25, 1985, had performed exceptionally well, placing a satellite in orbit. Everyone connected to the mission considered it well done, and exchanged congratulations. There was clearly a wind of truce blowing between the Air Force and NASA.

On February 14, 1985, a high-ranking meeting took place in the Old Executive Office building, Washington, DC, between the Under Secretary of the Air Force and the NASA Administrator that reached a truce between the two agencies. It said that NASA and DOD would continue to work together to ensure Shuttle is a viable operational system heading toward 24 flights annually.

In exchange for this agreement, NASA agreed with the need for a CELV, and supported the Air Force decision to buy ten and launch two each year from 1988 to 1992. Additionally, NASA /DOD agreed to work together on a second generation Space Transportation System, using manned and unmanned capabilities to meet all user needs, and a new pricing policy giving DOD incentives for its flights on the Shuttle.

The agreement was then incorporated into the National Security Decision Directive, known as the "National Launch Strategy," signed into law by the President on February 25, 1985. Although the House Subcommittee on Space Science and Applications called for the first public hearing, little was said during that meeting other than in praise of the agreement.

For its part, the Air Force seemed elated about the agreement, and with Congressional blessing, started the new CELV program.

The Titan 34 D7 was now called Titan IV and the Air Force was proceeding with plans to modify a small number of Titan II ICBMs and convert them into space boosters for satellites. The problem was, all the military satellites had been built to Shuttle specifications. Therefore, they were too large to fit CELVs.

When President Dwight D. Eisenhower signed the National Aeronautics and Space Act into law on July 29, 1958, he created our nation's space agency. It was an organization to "plan, direct, and conduct aeronautical and space activities," to "arrange for participation by the scientific community in planning scientific measurements and observations," and to "provide for the widest practicable and appropriate dissemination of information concerning its activities and the results thereof."

It was a space agency created to guide the nation into space through peaceful means. The original language did not include military use of space, or paying for programs that were used in space. Subsequent policies changed that, without formally changing NASA's original Act.

NASA had insisted that the Space Shuttle could launch all military satellites—even the heaviest. So military satellites were made a certain size and weight in order to fit into the Shuttle bay. Yet these same satellites were too heavy for ICBM- type boosters. Therefore, when Challenger exploded, the nation was without any capability to launch military or civilian spacecraft or satellites for nearly the next three years.

Appendix III – Blattner Annotation

1) George M. Low, Team Leader, NASA Transition Team, to Mr. Richard Fairbanks, Director, Transition Resources and Development Group, December 19, 1980, "Report of the Transition Team, National Aeronautics and Space Administration. NASA Act, Sec. 102. (a) The Congress hereby declares that it is the policy of the United States that activities in space should be devoted to peaceful purposes for the benefit of all mankind.

2) Preamble to the U.S. Constitution: We the People of the United States, in Order to form a more perfect Union, establish Justice, insure domestic Tranquility, provide for the common defence, promote the general Welfare, and secure the Blessings of Liberty to ourselves and our Posterity, do ordain and establish this Constitution for the United States of America.

3) On the Shoulders of Titans: A History of Project Gemini by Barton C. Hacker and Charles C. Alexander.

4) Les Blattner interviewed Edward C. "Pete" Aldridge, Jr. on October 20, 1999.

5) July 4, 1982, National Security Decision Directive Number 42 National Space Policy. The directive established national policy to guide the conduct of United States space program and related activities; it superseded Presidential Directives 37, 42, and 54, as well as National Security Decision Directive 8. The directive was consistent with and augments the guidance contained in existing directives, executive orders, and law. The decisions outlined in the directive provide the broad framework and the basis for the commitments necessary to conduct United States space programs.

6) Verified by Edward C. Aldridge, Jr. who attended the meeting, and was consulted when Secretary Weinberg sent the letter to the President.

APPENDIX IV

ENGINEERING DRAWINGS OF THE SHUTTLE AND HISTORY OF SHUTTLE O-RING FAILURES

NASA and Morton Thiokol drawings ("PC" numbers on documents are those referenced in the U.S. National Archives, the records available to all U.S. Citizens.)

1. The Space Shuttle System

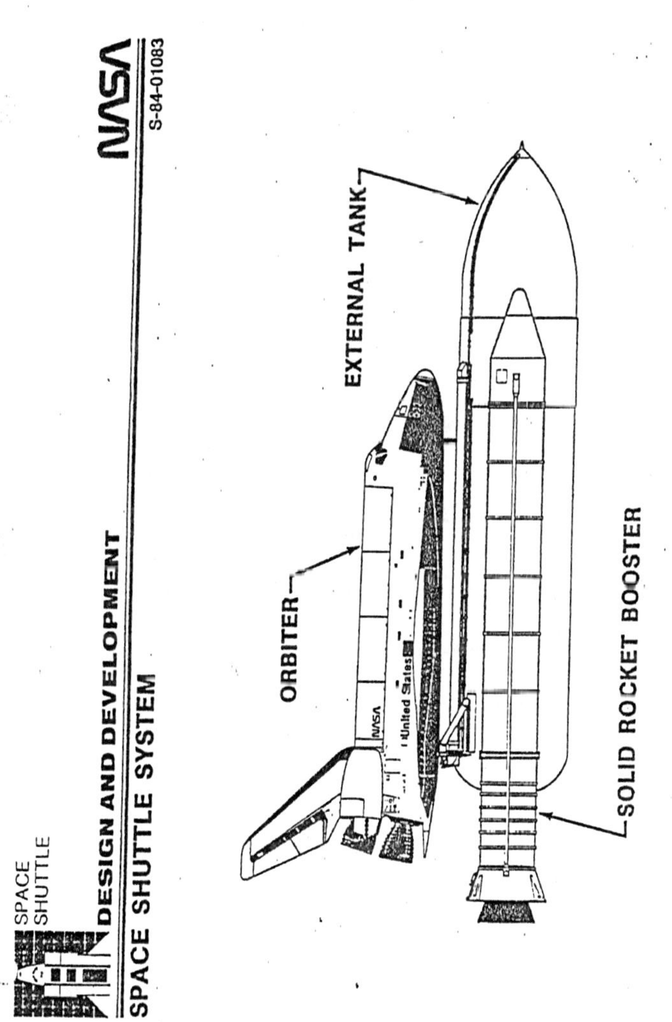

2. Orbiter Mechanical Subsystems

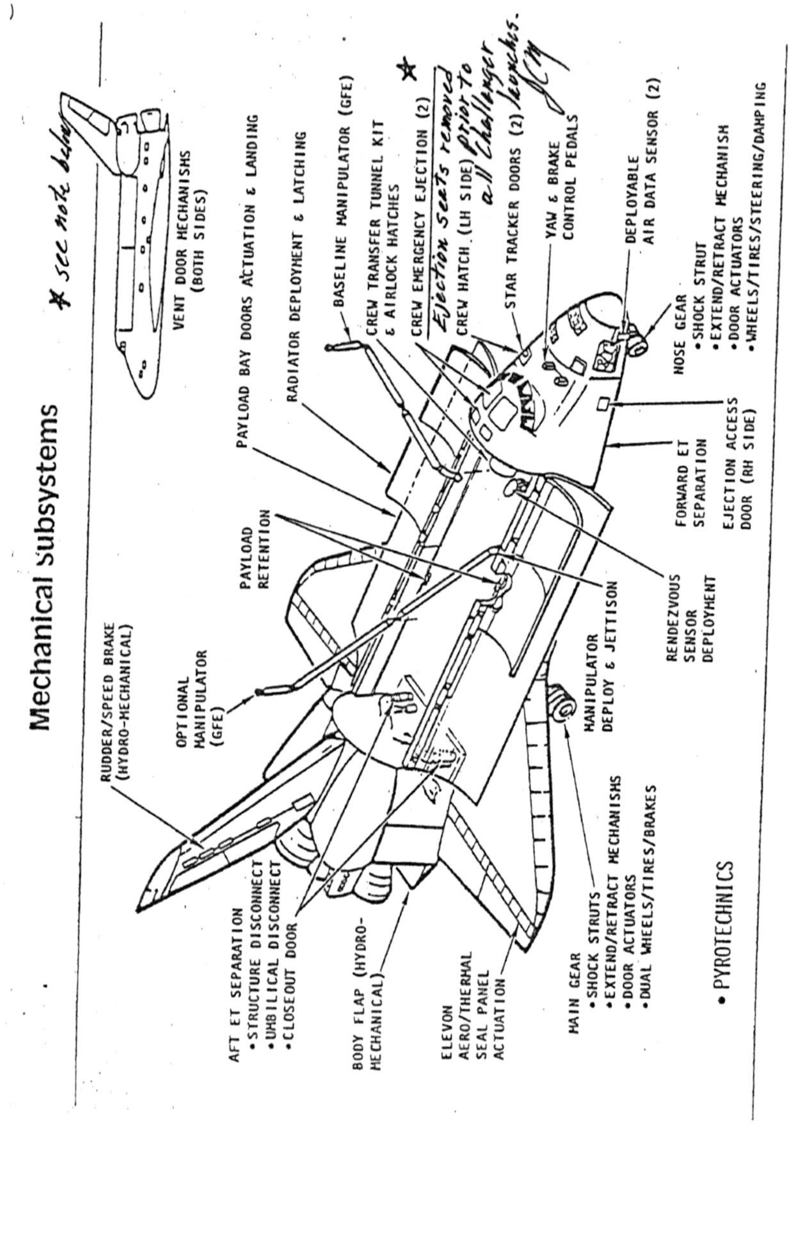

3. Space Shuttle Spacecraft Structures

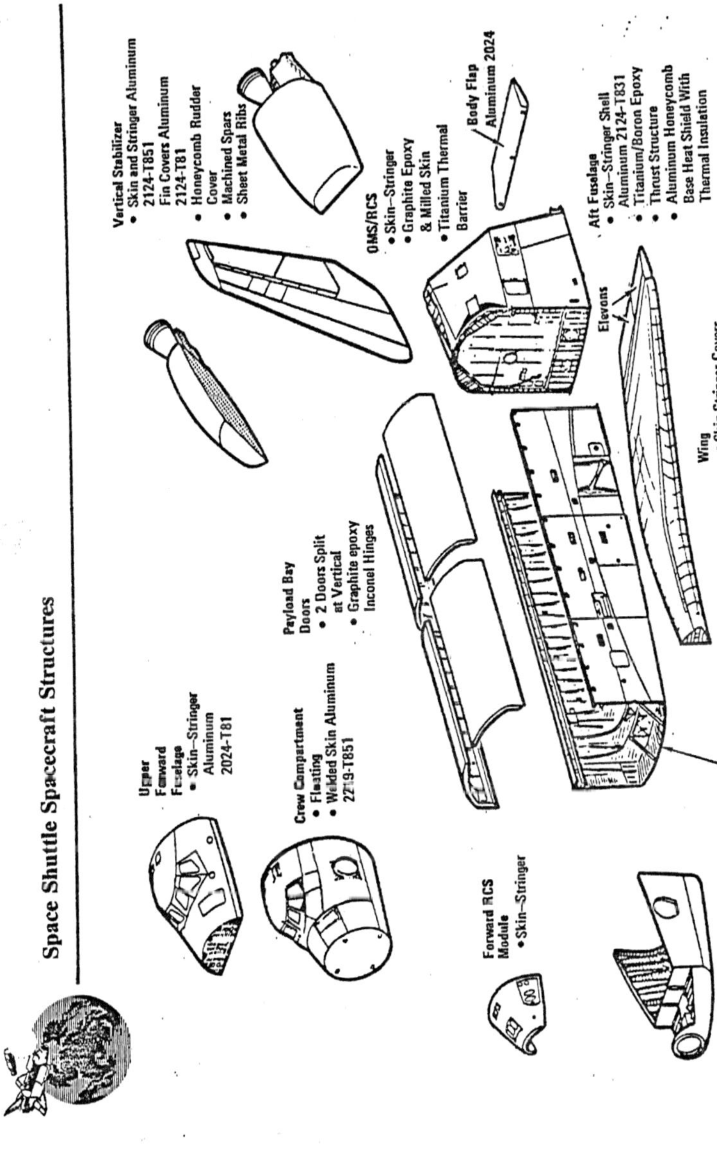

4. Crew Cabin Arrangement and Crew Functions

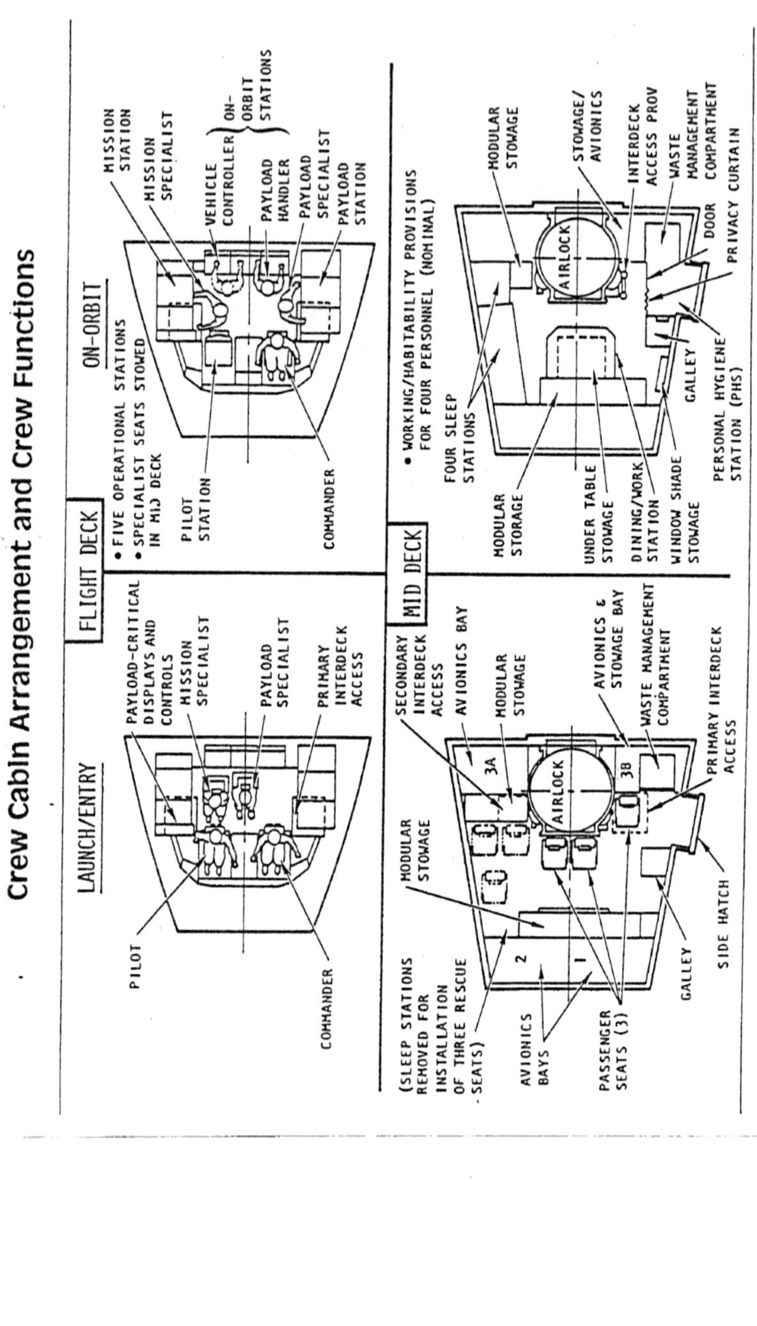

5. Measured Shuttle Temperatures on the morning of the accident launch.

MEASURED SURFACE TEMPERATURES – OMEGA SCOPE

January 28, 1986 Between 7:20 and 8:40 EST

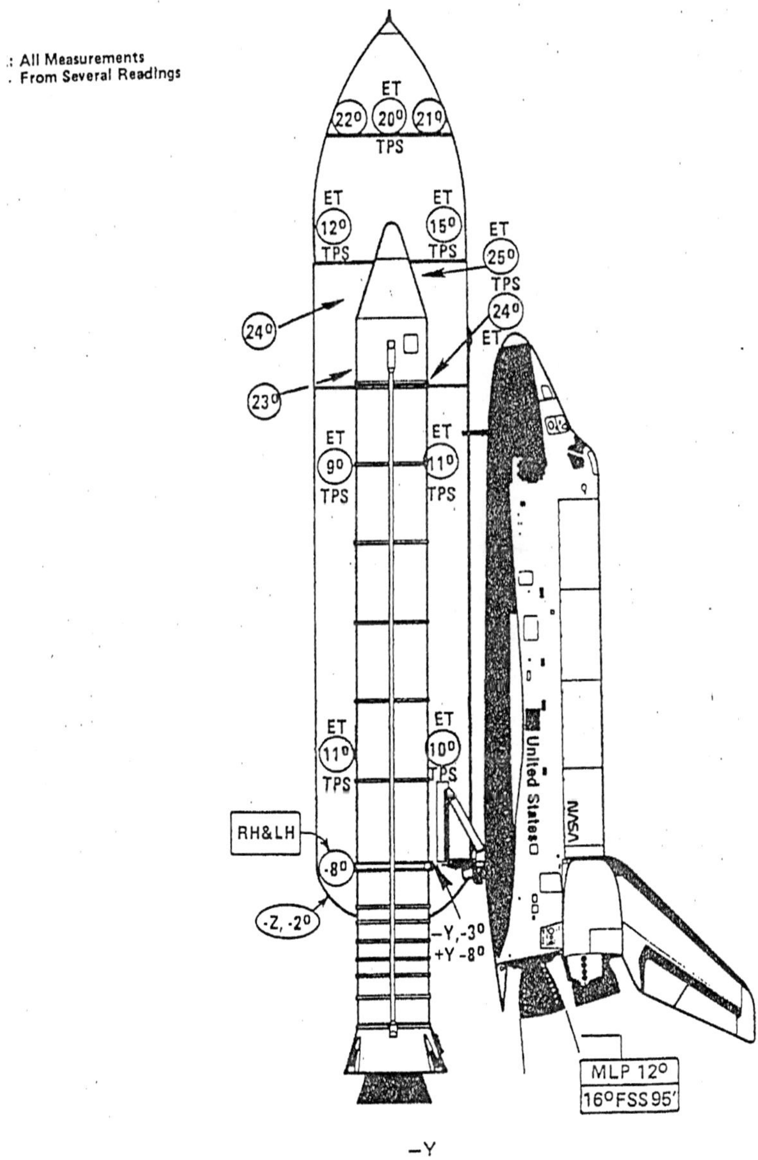

6. History of O Ring Damage in Field Joints - 2 pages

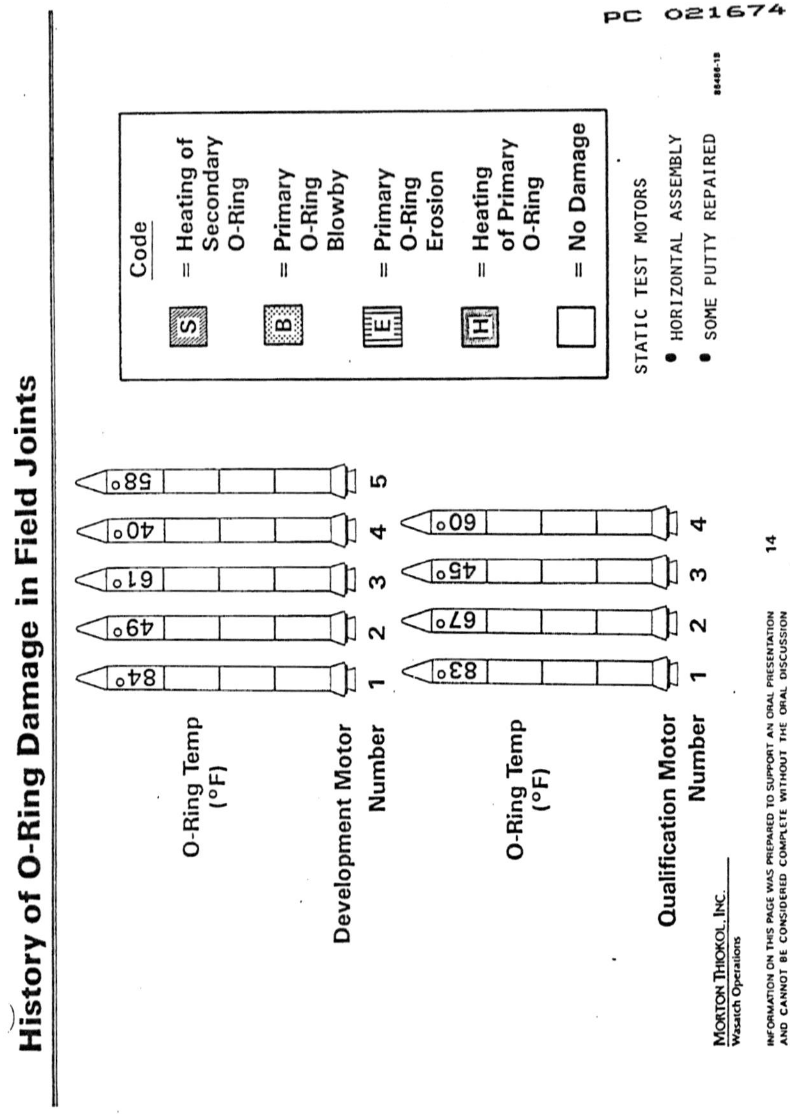

History of O-Ring Damage in Field Joints (Cont)

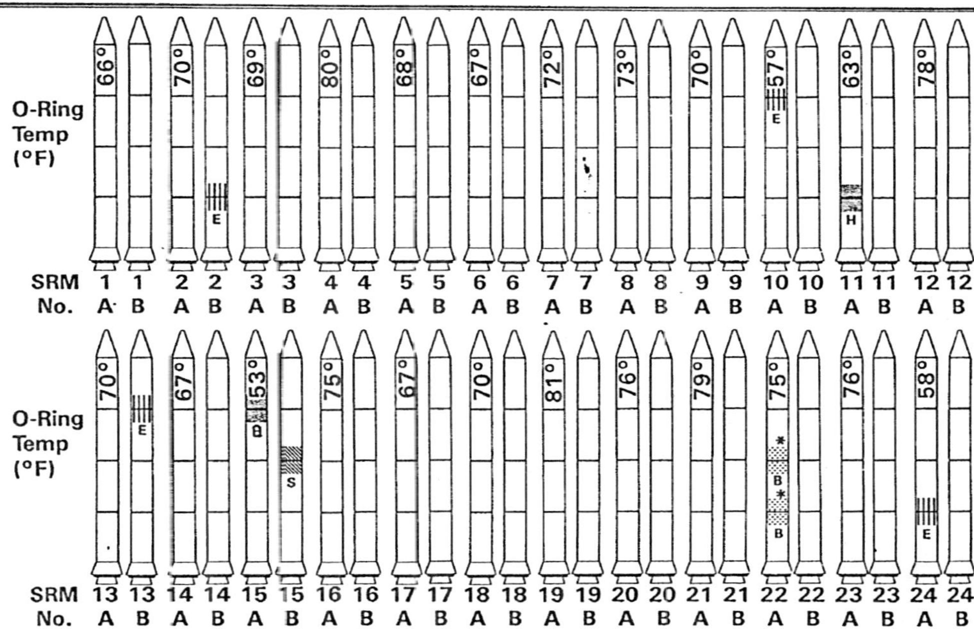

Morton Thiokol, Inc.
Wasatch Operations

* No Erosion

INFORMATION ON THIS PAGE WAS PREPARED TO SUPPORT AN ORAL PRESENTATION AND CANNOT BE CONSIDERED COMPLETE WITHOUT THE ORAL DISCUSSION

15

PC 021675

APPENDIX V

PHOTOGRAPHS

NASA HEADQUARTERS, CAPE CANAVERAL, FLORIDA – March 7, 1986

Left to right, Dr. Alton Keel, David Acheson, Astronaut Robert Overmeyer, Jack Macidull, and Chairman William Rogers. Signed by Rogers, "To Jack Macidull with best wishes."

MORTON THIOKOL, OGDEN UTAH – March 19, 1986

Left to right, Ron Waite, General Kutyna, Jack Macidull, and Neil Armstrong. Signed by Armstrong, "Jack, Sincere thanks for all your valuable contributions to the Presidential Commission."

WHITE HOUSE ROSE GARDEN CEREMONY TO PRESENT THE FINAL WHITE HOUSE COMMISSION REPORT – June 6, 1986

Challenger Commission with President Reagan. Signed, "To John Macidull with best wishes, Ronald Reagan."